D1345907

Small Farm
Mechanization
for
Developing
Countries

Small Farm Mechanization for Developing Countries

Peter Crossley
and
John Kilgour
Silsoe College, Cranfield Institute of Technology, Bedford, UK
with a chapter by
J. Morris

JOHN WILEY & SONS
Chichester·New York·Brisbane·Toronto·Singapore

Library of Congress Cataloging in Publication Data:

Crossley, C. Peter.
    Small farm mechanization for developing countries

    Includes index.
    1. Underdeveloped areas—Farm mechanization—
Handbooks, manuals, etc. 2. Underdeveloped areas—
Farms, Small—Handbooks, manuals, etc.   3. Farm
mechanization—Handbooks, manuals, etc.   4. Farms,
Small—Handbooks, manuals, etc.   I. Kilgour,
John.   II. Title.
S760.5.C76 1983   631.3   83-5935

ISBN 0 471 90101 6

British Library Cataloguing in Publication Data:

Crossley, C. Peter
    Small farm mechanization for developing countries
    1. Farms, small—Underdeveloped areas
    2. Farm mechanization
    I. Title   II. Kilgour, John
    338.6'42   HD1476

√ISBN 0 471 90101 6

Photo Typeset by Macmillan India Ltd., Bangalore

and printed in Great Britain by Page Bros (Norwich) Ltd.

Contents

vi

Preface

This book has been written by two agricultural engineers, with very useful assistance (in Chapter 8) from an agricultural economist. We have all been involved with the problems of small farm mechanization for a number of years. Our aim has been to use engineering and economics respectively as tools to tackle the particular problems of agriculture in the small farms sector of developing countries. The aim is continued in this book, which should be regarded as an aid to making and implementing the basic decisions relating to power in agricultural mechanization.

Since ours is a physical world, the application of physical science (or engineering) to the solution of its problems will always be necessary and relevant. But so, too, are considerations of economics and people. Even if the engineering is 'right' there is no guarantee that a machine or system will be regarded as socially acceptable or be economically viable. This point is often made and is a very valid one. The argument in favour of tackling the engineering, however, is that a machine must be able to perform its basic functions before it can even begin to be regarded as economically or socially acceptable. That is why we have endeavoured to show here how engineering principles and techniques can be applied to the particular problems of smallholder mechanization.

UNITS AND THEIR USE

Throughout the book examples are given within the text and, in order to increase their usefulness, realistic values are inserted wherever possible. The units used are SI (i.e. metric), as opposed to the traditional 'Imperial' units.

SI Units and Their Use

Quantity	Unit	Abbreviation	Comparison (approx.)
Mass	kilogramme	kg	2.2 lb
Length	metre	m	3.3 ft or 1.1 yds
Time	second	s	
Temperature	degree Celcius (or Centigrade)	°C	

The above are 'basic' units. From these, other useful units can be derived:

Quantity	Unit	Abbreviation	Comparison (approx.)	
Force	newton	N ($= kgm/s^2$)	0.22 pounds force (lb f)	
	kilonewton	kN ($= 1000\,N$)	225 pounds force	
Work or energy	joule	J ($= Nm$)		
Power	watt	W ($= J/s$ or Nm/s)		
	kilowatt	kW ($= 1000\,W$)	1.34 horsepower	✓
Area	sq. metre	m^2	10.76 sq. ft	
	hectare	ha ($= 10{,}000\,m^2$)	2.47 acres	✓
Volume	cubic metre	m^3	35.3 cubic ft	
	litre	l (10^{-3} cubic metres)	0.22 gallons	✓
Distance	kilometre	km ($= 1000\,m$)	0.62 miles	✓
Torque	newton metres	Nm	0.73 lbf ft	
Length	millimetre	mm (10^{-3} metres)	0.04 inches	

Further derived units, such as pressure (N/m^2) and flow (m^3/s) can be produced from combinations of the above. The useful feature of SI units is the 'coherence', that is the ability to combine into simple relationships with consistent units. For example:

Drawbar power (W) = Drawbar force (N) × Forward speed (m/s)

This can be checked as follows:

Left-hand side units are W or Nm/s, which equals right-hand side units of N × m/s. (It should always be possible to balance the units of an equation in this way.)

MULTIPLES AND FRACTIONS

When the basic SI unit is an inconvenient size for a particular measurement, the measurement may be specified in terms of multiples of fractions of the basic unit:

Multiplying factor	Prefix	Symbol
1,000,000,000	giga	G
1,000,000	mega	M
1,000	kilo	k
0.01	centi*	c
0.001	milli	m
0.000,001	micro	μ
0.000,000,001	nano	n
0.000,000,000,001	pico	p

* the fraction 1/100 is traditionally used in the unit of length (centimetre) but should be avoided in combination with other units.

In general measurements should be specified in terms of units such that the 'whole number' part of the measurements is between 1 and 10,000 thus:

300 mm *not* 0.30 m

90 kN/m² *not* 90 000 N/m²

CONVERSION FACTORS

Conversion factors for some of the more common units are given in the table below, where Imperial Units x k_1 = SI units and SI units x k_2 = Imperial Units.

Imperial	k_1	SI	k_2	Imperial
Length				
in	25.4	mm	0.0394	in
ft	0.305	m	3.28	ft
yd	0.914	m	1.09	yd
mile	1.61	km	0.621	mile
Area				
in²	645	mm²	0.00155	in²
ft²	0.0929	m²	10.8	ft²
yd²	0.836	m²	1.20	yd²
mile²	2.59	km²	0.386	mile²
acre	0.405	ha	2.47	acre
Volume				
pint	0.568	litre	1.76	pint
gal	4.54	litre	0.220	gal
US gal	3.79	litre	0.264	US gal
bushel	36.4	litre	0.0275	bushel

Mass

oz	28.4	g	0.0353	oz
lb	0.454	kg	2.204	lb

Density

lb/in³	27.7	g/cm³	0.0361	lb/in³
lb/ft³	16.0	kg/m³	0.0624	lb/ft³

Force

lbf	4.45	N	0.225	lbf
tonf	9.96	kN	0.100	tonf

Pressure

lbf/in²	6.89	kN/m² (kPa)	0.145	lbf/in²

Energy

ft-lbf	1.36	J	0.737	ft-lbf

Torque

lbf-ft	1.36	Nm	0.737	lbf-ft

Power

hp	0.746	kW	1.34	hp

RULE OF THUMB CONVERSIONS

A few easily remembered relationships between SI and Imperial Units will enable a large proportion of everyday conversions to be made with reasonable accuracy. The following relationship may be found useful:

	Relationship		*Accuracy* %
Length	2 in	= 5 cm	$1\frac{1}{2}$
	5 miles	= 8 km	$\frac{1}{2}$
	100 yd	= 90 m	$1\frac{1}{2}$
Area	5 acres	= 2 ha	1
Volume	7 pints	= 4 litres	1
	2 gal	= 9 litres	1
Mass	11 lb	= 5 kg	$\frac{1}{4}$
	1 cwt	= 50 kg	$1\frac{1}{2}$
	1 ton	= 1000 kg (1 tonne)	$1\frac{1}{2}$
Force	2 lbf	= 9 N	$1\frac{1}{4}$
	1 tonf	= 10 kN	$\frac{1}{2}$
Pressure	15 lbf/in²	= 100 kN/m²	3
Yield	8 cwt/acre	= 1000 kg/ha	$\frac{1}{4}$
		= 1 tonne/ha	$\frac{1}{4}$
Power	4 hp	= 3 kW	$\frac{1}{2}$

Introduction

Smallholder agriculture plays a very important part in most developing countries, even in those with other resources such as minerals or fossil fuel. A major reason is that a large proportion of the population (often in the range 60–70 per cent) will usually be engaged, either partly or wholly, in this sector of agriculture. In many cases the farming will be of the subsistence type, which is therefore vital in providing food for the majority of the rural population. Those farmers producing excess produce will be able to trade or sell it into either the local infrastructure or the urban markets and, in the case of some commodities, into an export marketing system. In all cases this constitutes a very important input into the economy of the country.

Because of growing urban populations and other factors, a number of developing countries which were originally net exporters of staple foodstuffs are now net importers. Where the foreign exchange for such imports has been gained from high-value export crops grown in place of the staple produce, this can represent a satisfactory situation. Such, however, is often not the case. There remains in most developing countries an important need to increase productivity, both of staple foodstuffs and of export crops, at the smallholder level. In order to do this it is usually necessary either to increase cultivated area or to improve on (effective) yields per unit area through better inputs, better management, and better storage and processing.

In many situations significant increases of the types mentioned will involve increased power inputs for such operations as primary cultivation, weed control, harvesting, and transport. The question then becomes—where is the increased power input to come from? The answer will usually be from one, or a combination of, the three basic alternatives: hand labour, draught animals, or engine power.

The decision as to which of these three is the most suitable will vary with the situation. It is not the intention of this book to suggest for example that large-scale, engine-powered mechanization is necessarily the answer to the problems of world agriculture. Many will argue that this approach has been tried and found wanting. Nor is it suggested that small-scale, engine-powered mechanization is inevitably the solution. Again, many attempts in this area have been tried without resounding success.

—— The approach adopted in this book is that, in a given farming situation, there are certain tasks that need to be carried out in order to produce a crop, and certain conditions exist which will affect the way these tasks can be carried out. It describes the characteristics of a number of alternative ways of performing the tasks, and aims to familiarize the reader with the engineering principles required in order to be able to analyse what is happening, to choose between alternatives, and to improve the performance of the alternatives where it is lacking.

—— In the final analysis, the choice of power alternative will still rest between the basic three (hand, animal, engines). In view of the difficulties associated with labour displacement, foreign exchange for equipment purchase and fuel, and repairs and service problems for engine-powered technology it is likely that the basic three will (and should) be considered in the order stated. In other words, hand labour is simple and flexible and should be adopted where possible, *unless* there is a requirement for more power, in which case draught animals may be considered, *unless* disease, fodder or other restrictions inhibit their effective use, in which case suitable engine-powered equipment may be an answer.

Part 1 *The Agricultural Requirement*

1 *Field Tasks and Conditions*

Typical smallholder conditions in East Africa—tree crops (coconut and cashew) mixed with maize and cassava

This chapter looks at the range of operations which are often required at the smallholder level and at the conditions in which the operations will be performed. Since the terms 'operation' and 'task' tend to imply the use of equipment and techniques in some combination, it is difficult to avoid mentioning some of the types of equipment which may be used. The intention, where equipment is mentioned in this chapter, is for it to be treated as examples of the kind of performance characteristics which are normally associated with given tasks. For example, it is not easy to discuss chemical weed control without mentioning sprayers. More detailed performance characteristics of various sprayers are given in Chapter 7. In the present chapter they are introduced in order to point out the possibilities. Throughout this chapter the emphasis should be placed on two questions, firstly, 'What is required to get the crop to grow and then to get it in its required form to the farmstead?' (i.e. *Tasks*) and secondly, 'What kind of factors are we likely to become involved with while doing this, and what effects may they have on the equipment we are liable to be using?' (i.e. *Conditions*).

1.1 TASKS

1.1.1 Primary cultivation

The object of primary cultivation is to disturb the soil in such a way that it can be subsequently processed into a condition suitable for germination of the seed when planted. Depending on the circumstances there are therefore a number of reasons why primary cultivation may be carried out. These are:

(1) To control weeds by inverting the soil and burying them.
(2) To increase infiltration and reduce run-off of water received from natural precipitation or irrigation.
(3) To disturb the soil at depth so as to allow the plant roots to penetrate deeper. This may increase the amount of water available to the plant, particularly in short-term drought conditions.

It is evident that these factors will not be required in all conditions. Indeed the action of inverting the soil in dry conditions may not be beneficial to weed control and may increase the rate of water loss from the soil surface through evaporation.

The traditional implement for primary cultivation in temperate regions is the mouldboard plough, which will achieve objective (1) above admirably. An alternative is the disc plough which, when suitably set, will give similar results. Neither of these has a significant effect in disturbing the soil below ploughing depth, and indeed it is likely that in some conditions the practice of ploughing with one tractor wheel in the furrow will lead to smearing and increased resistance to root penetration.

A chisel plough or tine is useful for disturbing the soil without permitting smearing from the tractor wheel. No inverting is achieved, and weeds must therefore be controlled by other means. In semi-arid conditions, soil inversion is not in any case beneficial and weeds can often better be destroyed by burning. Where the soil is fairly hard it will tend to shatter into a V shape at approximately 45° either side of a narrow tine. Chisel ploughing at close intervals, therefore, causes the fractured areas to break into each other to give a satisfactory overall soil disturbance. It has been found, for example, that by drawing a single tine through hard, dry soil at a depth of 150 mm and a spacing of 300 mm, the resultant disturbance is sufficient to allow a secondary operation such as ridging to be carried out without further cultivation.

Chisel ploughs do not work well in non-friable soils such as moist clay, where they tend to cut a slot. Their performance in almost any conditions can, however, be improved considerably by the addition of 'wings'. (Figure 1.1). Considerable work on this aspect has been carried out at the National College of Agricultural Engineering, Silsoe, Bedford, UK (Spoor and Godwin, 1978).

Tests in Southern Africa (Willcocks, 1981) have shown that measures such as precision strip tillage, using a tine at 0.7 metre intervals in ferruginous soils, produced sufficient soil disturbance to allow root growth, increased the amount of water infiltration in the vicinity of the crop and required about half the energy

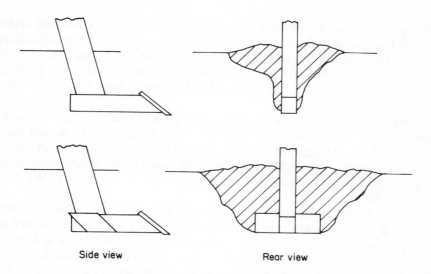

Side view   Rear view

FIGURE 1.1 Soil disturbance resulting from a subsoiler (upper view) and a subsoiler fitted with wings (lower view) after Godwin and Spoor (1978)

per unit area demanded by mouldboard ploughing. Similarly, the control of weeds in vertisols is a major reason for tillage, and this can be achieved using wide-level discs, resulting in less than half the energy requirements of chisel ploughing.

A further advantage of reduced or shallow tillage, where soil conditions allow it to be practised, is that the rate of work will usually be substantially higher than with conventional methods, reducing the time lost at the beginning of the growing season. This is particularly important where two crops per year are possible, and is as valid at the smallholder level as it is in large-scale operations.

1.1.2 Secondary cultivation

This is the operation or operations required to turn the soil into a seedbed suitable for planting. In some conditions this may require a number of operations using such implements as spring-tined cultivators, harrows, and rollers. It is felt in some quarters that this multiplicity of operations, particularly when carried out by heavy, wheeled equipment, does almost more harm than good to the soil, and efforts have been directed towards reducing the effects of repeated compactions from the tractor wheels. This can be achieved by such measures as 'ganging' implements, that is pulling a train of implements one behind the other, or indeed by reducing to a minimum (even to zero) the amount of tillage carried out.

It is argued that by disturbing only the area of soil in which the plant will actually grow (minimum tillage) or by not disturbing the soil at all and planting directly into it (zero tillage) the amount of damage done to the soil through compaction by the tractor wheels is minimized. A further technique aimed at this end is that of 'tramlines', where permanent wheel lanes are left in the soil and are used in all subsequent operations. This ensures that only the wheel lanes are run over by the tractor, and the lack of output from these areas is said to be more than balanced in most cases by the increased yields arising from the non-wheeled remainder.

In wetland cultivation such as paddy growing the churning action of draught animal hooves can be a useful input to assist in producing the optimum conditions required for growth of the rice plant. Here the use of heavy machinery can be undesirable, not only because of traction problems but also because of the danger of breaking through the hardpan established below growing depth to retain the water.

Where established structures such as ridges or beds exist, part of the secondary cultivation operations will usually consist of rebuilding the structure following the previous harvest. A 'wide bed' system advocated by ICRISAT in India is intended to combine water retention with flood prevention: beds are formed at 1.5 m centres, having a saucer-shaped cross-section as shown in Figure 1.2 (Binswanger, Chodake, and Theirstein, 1979).

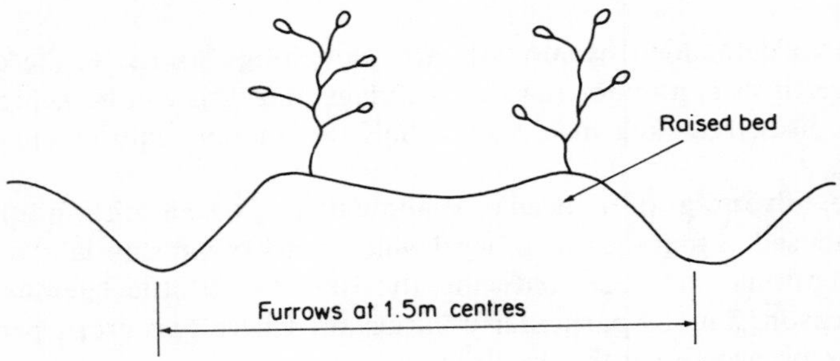

FIGURE 1.2 ICRISAT wide-bed system

Two rows are planted in each bed, the shallow dip in the centre of the ridge increasing water retention in dryer conditions, while the deeper furrows either side of the bed allow excess water to stand without waterlogging the crop. Cultivation in this case is a matter of rebuilding the bed as necessary, and drawing shallow tines along the areas to be occupied by the crop. The bed structure and position remain from one season to another and cultivation is based on a pair of draught animals walking along the furrows, pulling a wheeled toolbar on which the implements are mounted (Kemp, 1980).

Similar operations may be required where cultivation is performed on ridges, particularly where hand cultivation is normal. Ridges may be rebuilt by scooping material up from the furrow by the use of a ridger or by hand tools. A more suitable technique agronomically calls for a chisel plough to be drawn along the furrow to break any pan formed during the previous year, followed by splitting the ridges on to the previous furrows (Figure 1.3). This produces ridges having

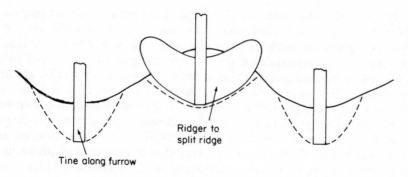

FIGURE 1.3 Ridge-splitting operation

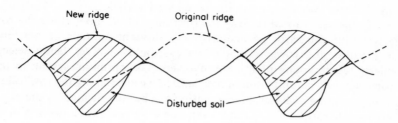

FIGURE 1.4 Disturbed soil in new ridges

considerably loosened soil within them (Figure 1.4), but requires high draught forces. It is not easy to steer a ridger during this operation as it tends to break out from one side or the other. This operation is more properly considered as primary cultivation.

1.1.3 Other operations

(i) Planting. This may be done either in rows or broadcast, depending on the crop and conditions existing. Generally, machine operations will require row planting, as will hand-based operations where ridges or other structures are used. The details of the operation will vary depending on seed size and other factors: large seeds will tend to be placed individually (or in twos and threes) in prepared holes; small seeds may be planted into a surface tilth and covered over. With the no-till system investigated extensively at IITA in Nigeria (Okigbo, 1981) and advocated for the humid tropics, the seeds are jab-planted into a surface mulch of vegetation which has been controlled by chemicals. The mulch is said to reduce soil erosion during heavy rain, and to retain the soil in a condition into which seeds can be directly drilled by the jab-planting method. This system is a modernized, row-based version of traditional agricultural practice in which a small space is cleared in the weed cover to allow a seed to be planted in a shallow hole.

(ii) Weeding. Weeding operations are one of the most important bottlenecks occurring in smallholder conditions, particularly where mechanization of the cultivation operations has allowed an increased area of land to be planted. The high temperatures and rainfall experienced in many areas cause rapid growth, not only of the planted crop but also of the weeds, which can quickly choke the former unless controlled. The use of chemicals for weed control has some advantages but these are probably reduced in the smallholder situation, where the problems of safety and the high cost of chemicals become more significant. Where chemicals can be used safely and effectively, as in a well-organized no-till system, they will probably offer a useful solution to weed control problems in a number of situations. Otherwise a mechanical system is required. This can make a very high demand on labour availability, and lack of the ability to control weeds effectively may significantly reduce the gains otherwise made from planting increased areas or using improved seed varieties.

Where crops are planted in rows the operation of cleaning along the row is relatively straightforward, particularly with draught animals or machinery. Weeding within the rows is much less easy and involves being selective, particularly where young plants are involved. The optimum system in many situations may be a combination of mechanized weeding along the row followed by intrarow clearing by hand. The operation can be complicated by the practice of interrow cultivation, where more than one crop occupies a given plot.

(iii) Spraying. This may be carried out for other reasons than weed control (e.g. insecticides, fungicides, molluscicides, nematicides, etc.). Sprayers are necessary for the application of these chemicals, and will either be hand-held (knapsack-sprayer type) or machine-mounted. All the types need to be controlled carefully to avoid overlap or missed areas, to minimize drift, and particularly to avoid safety hazards for the operator.

A problem with conventional sprayers in the smallholder situation is that large quantities of water are required which may not be readily available and will in any case require considerable energy expenditure for transport to the field. Ultra-low volume sprays now available use spinning discs or other devices which ensure that the droplets are very small and of roughly the same size and this allows very much lower quantities of water to be used (20 l/ha rather than 200 l/ha). By electrostatic charging the droplets are attracted to the plants and drift is further reduced (Bell, 1981).

(iv) Harvesting. It is of course necessary to harvest the crop once it is fully grown in order that it can proceed on its way along the food supply chain (whether to the farm, village, local market town, capital city, or the world export market). The object of a harvesting operation is to obtain the maximum possible proportion of the crop, in a suitable condition and with the minimum of input. This is not easy to achieve. A certain proportion will inevitably remain in the field, and where uneven ripening, lodging, rodents, birds, or delayed harvesting occur, this proportion can reach unacceptably high levels.

In most cases a plant consists of a vegetation part and a food part (in the form of grain, nuts, roots, or pods). Usually, but not always, the vegetation is either disposed of or used for other purposes such as bedding, building, fuel, etc. Harvesting therefore often involves separating the food portion from the vegetation portion, either in the field or elsewhere. Separation in the field is used for some crops such as maize, where the vegetation part of the plant is often left in its original position or flattened. However, the majority of crops will involve a cutting operation followed by a separation process, often after a further period of drying in the field. Rice is an example of an above-ground crop which is first cut and then separated; groundnuts is an example of a below-ground crop, where the cutting operation is replaced by a soil-disturbing and lifting process.

It is important, when considering equipment and techniques for harvesting, to be in a position to decide what cutting/lifting and separation operations are necessary and when they *could* be undertaken. Some conventional methods may be used because of a lack of new ideas or equipment. For example, dryland rice is usually cut and then separated from the stalk. It is perfectly feasible, however, to thrash the rice from the stalk in the field. This may require a change in the system for disposing of or utilizing the stalks, but the actual process can work well. Work at NCAE Silsoe some years ago resulted in a prototype small-scale rice harvester which separated the heads from the stalks in this way.

Consideration of this machine leads to another factor worth discussing—the time available for harvesting. This is often a constraint on production in areas where two or three crops a year are possible agronomically. In order to clear the land for cultivation it may be necessary to harvest the crop rapidly. The time available for subsequent cleaning and processing at the farmstead may then be much more extensive and it is therefore worth considering an approach which can result in a rapid harvesting process producing a crop which is not particularly well sorted. The rice harvester referred to above, for example, being simple, produced a harvested crop which contained quite a large proportion (perhaps 10 per cent) of vegetation and rubbish. However, this could be transported quickly to the farmstead and sorted later, clearing the field for further cultivation and planting.

1.1.4 Energy considerations

To draw an implement through hard, dry soil typically requires a pull of up to 5 kN. Assuming that the effective implement width is 0.33 metres and that the implement moves at 1 m/s (thereby requiring a power of 5 kW at the implement), it would take 8.33 hours to cultivate a hectare, resulting in an energy requirement of 5×8.33 kWh or 41.7 kWh per hectare. This is equal to 150 MJ (megajoules) or 150 million joules (1 kWh = 3.6 MJ). It can thus be seen that a great deal of energy can be required in order to carry out a primary cultivation task such as that described. (The relationships used above will be derived in later chapters.) Practical results give over 100 MJ/ha for primary cultivation (Willcocks, 1981) and about 40 MJ/ha for secondary tasks.

In softer conditions the energy required will be lower, but in total with a number of operations the energy will be of a similar magnitude and this energy must be provided in order to carry out the operations. The source of the energy (human, draught animal or mechanical means) may vary and may provide the energy at different rates (power) and efficiencies. (It should be noted that the above energy figures are those required by the implement.)

As an example, using a conventional tractor the overall efficiency may be, say, 65 per cent which means that 64 kWh of energy per hectare may be required from the engine in order to provide 41.7 kWh at the implement. A tractor diesel engine will typically consume fuel at the rate of 0.35 l/kWh, so that about 22 litres of fuel will be used per hectare.

Calculations for human or draught animals are not so straightforward, as both consume energy (in the form of food) for working and living, and the energy input is not related directly to the output in the way that it is with a machine. Nonetheless it is evident that, whatever the source of the energy and whatever its power, the same basic amount of energy must be provided by one means or another (Inns, 1980).

The amount of energy required per hectare can be reduced by waiting until the soil is soft enough to cultivate easily. If the onset of rains results in the required force per implement reducing to 1.5 kN, the energy needed becomes 12.5 kWh/ha. A pair of oxen producing 1 kW would then take 12.5 hours to cultivate a hectare (whereas in the dry season it would have taken 42 hours, assuming that they could provide a pull of 5 kN, which is not likely). Working for 4 hours per day on (say) a 5 ha holding means, however, that more than a fortnight is required to complete the primary cultivation; with some crops two week's delay after commencement of the rains may incur considerable yield penalties.

One alternative would be to use a small engine-powered unit of say 5 kW effective (drawbar) power which could produce the required dry season pull of 5 kN at 1 m/s. The time per hectare would be 8.33 hours (as previously derived), and the total time for a 5 ha holding would be 41.6 hours (or about 1 week's work). To provide 5 kW at the drawbar would, in the case of a small tractor, normally require an engine power of about 8 kW which, at 0.4 l/kWh, would consume 133 litres of fuel for the holding. The large tractor mentioned earlier would use about 112 litres of fuel (it is more efficient because of less wheelslip). Whichever mechanical device is used, a great deal of energy is required. It will be shown later that a small winch-based cultivation device can work as efficiently as a large tractor (but much more slowly).

Minimum tillage techniques can be applicable to semi-arid conditions (see Section 1.1.1). Cultivating at a spacing of 0.75 m instead of 0.33 m will reduce the energy required from 41.7 kWh/ha to 18.5, the fuel (for a conventional tractor) from 22 litres to 10 and the time (for a small tractor) from 8.3 hours to 3.7. It should be noted that all the above times are for the cultivating task only and do not include idle and turn-round times. In practice an extra 30% or so should be added for this.

1.1.5 Power and rate of work

Power is the rate of doing work. Work has the same units as energy (and, incidentally, as quantity of heat) and may be expressed in joules or kW hours, as seen above. Power is therefore expressed in joules per second which is called a watt (W). Since this unit is rather small the unit kilowatt (kW) is normally used (1 kW equals 1.34 horsepower).

Although, as discussed in Section 1.1.4, the *energy* required to perform a certain operation will be the same regardless of the source, the *power* available will determine how fast that energy can be provided, that is the rate of doing work. Thus a pair of draught animals with a combined power of perhaps 1 kW will, all things being equal, take five times as long as a 5 kW machine to cultivate a given area. (All things are not always equal, of course, since efficiencies can vary.)

The rate at which the available energy can be made available (i.e. the power) thus determines how long it will take to perform a given operation, and a fundamental problem in smallholder agriculture is that the amount of power available is often not high enough to enable the required operation to be performed within the desired time. This can apply particularly to operations such as primary cultivation, weeding, and harvesting. This is the reason that mechanical power is so often advocated for the smallholder secctor. Despite its limitations, engine-powered mechanical cultivation has the potential for supplying power at a different order of magnitude from human or animal sources.

1.2 CONDITIONS

The objective of this section is to identify and describe those physical factors which may exist in typical smallholder conditions, and to mention the kinds of problems which can then occur with the use of machinery in these conditions. To a certain extent solutions are suggested, but the intention is to draw attention to factors which may cause problems, rather than to propose solutions at this stage. Later chapters in the book describe more fully the characteristics of various equipment, and suggest modifications which may be made to overcome certain problems.

1.2.1 Physical conditions

(i) Soil properties. The soil types found in developing countries cover the normal range of agricultural soils found elsewhere, but there are differences in their characteristics related to climate and previous treatments.

The effect of climate is particularly marked when there is an extreme dry season, since the months without rain cause the soil to dry out and become in some cases exceedingly hard. This situation can occur for example in much of East and Central Africa and, even where the seasons are bimodal, it may cause 'normal' agricultural soils to dry out to the extent where it requires a force in the region of 5 kN to disturb the soil to a depth of 150 mm using a single tool. An appropriate tool to use in this case would be the tine or chisel plough, which shatters the soil without inverting it; disc ploughs are satisfactory but difficult to operate with small-scale equipment; mouldboard ploughs are not particularly suitable in these conditions. Some so-called 'cracking clays' can dry out to the extent where cultivation with all but very large equipment is impracticable and being relatively structureless they are also very difficult to handle when wet.

Many soils will be quite easy to cultivate when wet, but in this case the traction of wheeled machines will suffer. Additionally, it is often advisable from an agronomic point of view to prepare a seedbed before the onset of the rains so that early planting can occur. Delay at this stage can reduce yields considerably in a number of crops such as maize, sorghum, and cotton. The problem then becomes one either of cultivating hard, dry soil during the dry season, or of getting traction problems during the rains. Draught animals do not suffer from excessive wheelspin, but are often in poor condition at the end of the dry season due to lack of feed, and require a period of good fodder before being in a position to do much useful work. This gives rise to further delays at a critical time.

Many small farms are cultivated by hand, which tends to cause a handpan in the soil at a depth of 75 to 100 mm. Once this is broken up mechanically it becomes fairly easy to cultivate the soil, but the drawbar force required to do this may be high (perhaps up to 6 kN per tool).

A potential problem with any system of cultivation, but particularly with mechanical cultivation where crops are likely to be planted in rows, is soil erosion. Natural cover usually provides high resistance to erosion of soil by wind or rain. Traditional methods of cultivation, where crops may be planted into a natural cover, will do little to increase erosion; broadcast and intercropping methods, also traditional to hand cultivation, may produce some erosion until a cover is established. Mechanical cultivation, where dry friable soil may be exposed and kept exposed by mechanical weeding between rows, can produce severe soil erosion on slopes unless carried out along the contour.

There is, in many developing countries, an increasing pressure on land as populations increase at rates of 2–4 per cent per annum. Such pressure causes existing cultivated land to be cropped more intensively, and can also result in sloping or marginal land, previously existing satisfactorily under a range system, being brought under cultivation. This land is unlikely to produce useful crops for long, since unsuitable cultivation methods will often be used, and soil erosion may in the process ruin it even for its former use.

(ii) Trash. Crops and weeds grow rapidly during the wet season, due to high temperatures and ample water supply, and many tropical crops such as maize produce luxuriant vegetative matter, which is usually left after harvesting. This may later be burnt off but still may present an obstruction to a tool passing through the soil. A chisel plough, which exerts a bursting rather than a cutting force, is less likely to be affected by blocking than other types of implement and is therefore useful in these conditions.

(iii) Rocks, roots, and stumps. These can be classed as obstructions, which may be present in smallholder fields, depending on the thoroughness of the initial land clearing. Where large boulders are buried in the ground they will often be left there, and will form a hazard to mechanical cultivation equipment. Stumps and roots of previous trees will present similar problems—the shared characteristics of these obstacles is that they do not represent a problem for hand cultivation systems, where rows are not often used. When attempting to mechanize, however, it is essential to plant in rows and then disruption, inconvenience, and damage can be caused to machinery trying to operate in the conditions described above. A useful feature of small-scale equipment is that it may more easily be manoeuvred around such obstacles and, being low-powered, is more likely to stall on encountering them, rather than the implement getting damaged as usually would be the case with larger machinery. This is one of the reasons why contractors using large equipment frequently charge more per unit area to cultivate small holdings than for large farms.

(iv) Dust. In the dry season many agricultural soils give rise to a fine abrasive dust when disturbed. Unprotected parts of machinery with relative motion (either sliding or rotating) will quickly wear in these conditions, particularly if they are greased. Examples are diesel fuel pump racks, Bowden cables, lever pivots, plain bearings, brake drums, etc. The only really effective solution is to protect them from the dust by means of seals. The simple, rubber V-ring type is very effective for this purpose. Engine air inlets and crankcase breathers must have very good, well-maintained filters in these conditions since even a small leak in the air filter can damage an engine severely in a few hours. This aspect is discussed further in Chapter 5.

(v) Heat and altitude. The major effect of these ambient conditions is to reduce the power of engines by anything up to 20 per cent. (This point is also discussed in Chapter 5.) The high ambient temperatures found in many situations may effect the cooling of either water- or air-cooled engines. Removal of the thermostat may help with water-cooled engines. Air-cooled engines are more difficult to monitor as there is no warning (such as steam) before they seize up. Frequent attention should be paid to checking the cooling air inlet and outlet areas for excessive dust or vegetation blockage and the shrouds should be removed as part of the regular maintenance in order to clear any potential blockages occurring around the cooling fins.

(vi) Field size and access. Field sizes in smallholder agriculture are often very small (of the order of 0.1 to 0.5 ha). This may be due to physical constraints or to subdivision over the years. The result is that, the smaller the field, the higher proportion of its time a machine will spend turning at the end of the field and completing odd-shaped sections. If a machine spends 20 seconds turning at the end of a field 150 m long and operates at 1.5 m/s, it will plough for 100 seconds each length and will thus spend 17 per cent of its time turning round. If the field is

only 50 m long the machine will plough for 33 seconds and will spend nearly 40 per cent of its time turning round. This makes for very low field efficiency. It should be noted that, if a small machine works at the same forward speed and takes the same turn-round time as a large one, the field efficiency will be the same in each case, regardless of the working width of the implement.

Access to smallholder fields is often a problem, again because they were often originally organized on a hand-cultivation basis. It is easy for a person to jump a stream, pass between two trees or climb a steep bank in order to approach a field. A machine does not possess this versatility and obstructions may need to be removed before access can be gained. Perhaps the most difficult access to arrange is where only a footpath leads between other farmers' fields to the desired plot. In this case only a draught animal or a very narrow machine such as a small, single-axle tractor can gain access to the field. This, combined with the small size of holdings, means that total *working* hours per machine may be uneconomically low. Government tractor-hire schemes which attempt to provide smallholder ploughing services have found this problem in many areas. When using large-scale machinery the only realistic solution may be to plough several plots at once and reimpose the boundaries afterwards; but this can lead to difficult social problems. Similar problems can occur where the construction of an access road requires agricultural land to be taken.

(vii) Slopes. The majority of smallholder plots where mechanization would be attempted are probably on flat or gently sloping ground, since this is where the larger, better cleared fields will tend to be. However, there are several types of situation where operation on slopes can be required. A particular example would be where high-value cash crops such as tea or coffee are being grown. In this case there are several possible problems:

(1) Stability problems when working across the slope and when turning.
(2) Traction and power problems when working up the slope.
(3) Safety problems when moving down a slope without the stabilizing effect of an implement engaged in the soil.
(4) Control problems when working across the slope in crops.

The stability characteristics of various machine layouts are described in Chapter 9, and traction aspects are covered in Chapters 6 and 9.

(viii) Ridges. Even fairly shallow slopes may give rise to serious erosion when disturbed soil is exposed to the torrential rains which frequently occur near the beginning of the wet season. For this reason the practice encouraged in many areas is to make ridges along the contour so that run-off is reduced. This can also be effective in conserving water, particularly where rainfall is intensive but of low total quantity. In some areas crops such as maize are grown on ridges even where the land is fairly level. Ties may be made at intervals across the furrow to reduce run-off when the contours are not followed precisely.

Any device expected to work in the smallholder situation is likely to encounter ridges, which may give rise to two types of difficulty:

(1) Working along the ridge requires either the ridge spacings or the wheel spacings to be adjusted to match each other. A two- or four-wheel machine will need to straddle one ridge at least; in the case of three-wheel machines the minimum to be straddled is two ridges. In each case the ground clearance under the machine must be sufficient to clear the ridge and perhaps a young crop growing upon it.

(2) Typical smallholder fields do not have clear headlands and, when attempting to turn a small machine in ridges, severe difficulties of traction and control may be encountered. Ridges at 1 m spacing (which is typical for larger crops) will usually give an effective curvature in the furrow of about 0.7 m diameter. This is similar to the diameter of wheels fitted to much small equipment (eg 6.00 × 16 tyres or smaller) and consequently makes progress across the ridges difficult (Figure 1.5).

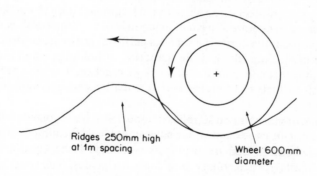

Ridges 250mm high
at 1m spacing

Wheel 600mm
diameter

FIGURE 1.5 A small wheel crossing ridges and furrows

(ix) Terraces. These structures exist on severely sloping land either where erosion could not be controlled by ridges or where better water retention and control is required for crops such as paddy. Again there are two major problems associated with the use of machinery in these conditions:

(1) Access is often very difficult, since it will inevitably be steep and probably narrow and winding. A human or a draught animal will find little difficulty in ascending a steep, twisting footpath to reach a terrace partway up a slope; a wheeled machine will encounter severe problems. A possible solution is to use a winch to haul the machine up the slope. Two NCAE prototype machines described later (Chapter 9) have built-in winches which could be used for this purpose.

(2) Although relatively level the cultivated areas are also often very narrow and irregular, making machinery operation difficult and inefficient.

1.2.2 Geographical conditions

Smallholder agriculture is very much a family concern and also provides the livelihoods of the majority of the population in most developing countries, either at the subsistence level or above. The result of these factors is that land tends to become increasingly subdivided as it passes from one generation to the next. The perceived need to share the risks of climate, disease, and pests more equitably between land-holding families means that the individual plots are often distributed over quite an extensive area, and the distances involved in moving between plots can be considerable.

This may not be a problem with hand or animal cultivation where work rate can be expressed in days per hectare rather than in hours. With medium and large-scale machinery, however, a large proportion of its time may be spent in moving from one plot to another (even where a number of farmers are being served, as with a contract hire system). This feature is therefore a significant one acting in favour of small-scale equipment, the low rate of work of which is often more equivalent to that of animal cultivation.

This clearly has an effect on costs, in that a large machine costing say five times as much per hour to operate as a small machine will be comparable economically only if it can work at five times the rate. Its performance characteristics may give it this possibility technically, but the scattering of holdings may mean that it is unable to attain this effective rate of work in practice. The cost per unit cultivated is therefore likely to be higher than for smaller, cheaper, and slower equipment.

The position of markets and supplies of inputs also has a considerable effect on smallholder agriculture, except where subsistence level farming with almost zero inputs is carried out. Indeed, the problems associated with increasing the level of agricultural inputs (seed, fertilizer, pesticides, herbicides, fuel, water, advice, and finance) must act as a deterrent to those seeking to move above the subsistence level and produce cash crops for sale in the external marketing system.

A considerable part of the problem is the condition of the feeder roads which must be used in order to connect with the input/output system, whether it be a local market town or a main road transport pickup point. As discussed in Chapter 3, the fact that feeder roads are normally of poor quality, particularly at certain times of the year, means that transport facilities are expensive and often difficult to obtain. Any machines or equipment intended to play a part in the transport system at the smallholder level (village to local market) will need to be robust, simple, and economical. Economy is easy to achieve if a vehicle has a reasonable capacity (500 to 1000 kg), and such vehicles do not need to be particularly fast, which can help to reduce their initial cost. (Studies have shown that in typical smallholder situations an economical type of vehicle is likely to be one which can carry about 750 kg at a speed of around 30 km/h, has good tractive performance without a complex transmission system, and is driven by a fairly low-powered—say 15 kW—diesel engine. Again, more information may be found in Chapter 3.)

1.2.3 Technical and social conditions

(i) Operator misuse. Misuse of a machine may be defined as its operation in a way which was not envisaged by the designer, and particularly in a way likely to cause damage to, or shorten the life of, components.

It may be argued that a good designer will accommodate possible misuse by anticipating for it. It is in practice difficult, however, to envisage and design for some types of operator misuse, since it may depend on: lack of awareness of the need for care in operation; lack of training in the detailed construction of the machine; lack of motivation to treat a machine reasonably (particularly evident with hired equipment); unusual (and severe) operating conditions; and lack of infrastructure or services causing the failure of certain components and leading to consequent increase of wear and tear on others.

Some examples will show the scale of possible misuse:

(1) Tractor batteries in poor condition due to lack of electrical test and service facilities, leading to the consistent necessity to tow start or 'bump start' tractors, with consequent transmission failures.
(2) Operation of a tractor in too high a gear while ploughing (black smoke is a giveaway sign).
(3) Poor clutch control (snatch engagement, noisy gear changes, 'riding' the clutch).
(4) 'Showy' operation (racing the engine, skid stops and wheel-spinning starts).
(5) Impact with rocks, trees, stumps, etc. due to poor steering control or lack of care.
(6) Overloading of vehicles with produce or passengers.

(ii) Servicing facilities and spares availability. The majority of the more complex machinery used in agriculture in developing countries will be imported, although some components (tyres, batteries, glass, sheet metal parts) may be supplied locally. Despite import restrictions on equipment, a wide range of different makes will often be available in many developing countries. This has advantages in principle in terms of competition but in practice it may result in too many companies chasing a relatively small market (in Kenya, for example, about twenty different makes of tractor have been imported in recent years). Each of the major established makes therefore has only a proportion of the market, and is continually 'harassed' by other firms trying to increase their own share, sometimes at the expense of the spares and service infrastructure.

Servicing is then carried out by agents, in many cases, rather than dealers, so that brand expertise and specialist knowledge are lacking. Certainly, the supply of well-trained machinery repair technicians in rural areas is unlikely to be anywhere near the required levels, as training is expensive and, when completed, tends to be rewarded more in the urban centres.

In some countries a considerable level of skill exists at the village artisan level, and equipment which is basically simple in construction will stand some chance of repair through this activity. The supply of spare parts, however, is a constant source of delay and expense. It may happen that machinery can be supplied in the first instance under an aid or loan agreement, but spares are the responsibility of the government. This involves the use of valuable foreign exchange, therefore it is likely to be restricted. At the commercial level a dealer or agent will find that obtaining spare parts from companies in another part of the world is expensive, and the capital tied up in stocks is considerable. This tends to result in stock levels which are far lower than desirable for efficient maintenance of the existing equipment, and the cost of spares when available may be four or five times their cost in the producer country (Inns, 1978).

This increases the cost of machine operation and, when spares are not available within a reasonable time, encourages the practice of 'cannabalization' where a tractor (having been laid up awaiting a part which is perhaps not particularly major) has another part removed from it to replace a faulty component on a second tractor. This renders the first tractor less viable, as it is now awaiting two parts, with the result that soon a third part is removed for the same reason. After this process has gone through a couple more cycles it is likely that secondary damage (for example corrosion of the bores following removal of a cylinder head) will make future use of the machine virtually impossible.

This is a thoroughly inefficient process, not only because a tractor consisting of literally hundreds of parts is now irreparable because of the lack of a handful of them, but also because it is unlikely that it will even function satisfactorily as a source of spares in the future. Neglecting the consequential deterioration mentioned above, it might be thought that eventually each part will be required to repair another tractor. In practice, of course, there are a considerable number of parts which will almost certainly never fail on other tractors and will therefore never be required; and conversely there are inevitably a number of parts, which due to a design weakness in the model or to specific operating condition hazards, will fail repeatedly. The cannabalization process does little to reduce this problem.

(iii) Social acceptability. It is a mistake to imagine that satisfactory function is the sole reason for a farmer buying and using a piece of equipment. A machine should perform its required tasks adequately but in addition it should impart a certain degree of satisfaction to its owner. This satisfaction can come from reduced drudgery, such as riding on a machine instead of walking behind it; and it can also come from the appreciation that the machine is the latest technology, equipped with all the available systems such as lights, starter, hydraulics, and trim. Any machine which is seen as simple, crude, slow, or old-fashioned may be rejected, particularly where the existence of up-to-date equipment on large-scale projects provides a 'norm' for comparison.

It is an unfortunate fact from an engineering point of view, however, that a machine incorporating additional systems such as electrics and hydraulics is much more likely to fail in marginal maintenance conditions. Spring starters or hand-starting, and manually-operated implement lifts are alternatives which are very unlikely to cause problems.

Experience with the design and testing of small-scale equipment in Africa has shown that, in the difficult operating conditions often encountered, a simple and robust machine is likely to be the most successful. It is, however, a more difficult matter to convince farmers exposed to comparisons with 'proper' tractors. It is to be hoped that effective, economic, and long-term operation of equipment, large or small, will gradually come to be regarded as the primary reason for purchase.

REFERENCES

Bell, R. D. (1981). The relevance of recent developments in pesticide application technology to crop production in the tropics. Fourth Session of FAO panel of experts on agricultural mechanization, Montpellier, 23–26 November.

Binswanger, H. P., Chodake, R. D., and Thierstein, G. E. (1979). Observation on the economics of tractors, bullocks and wheeled tool carriers in the semi-arid tropics of India. ICRISAT Workshop, 19–23 February.

Inns, F. M. (1978). Operational aspects of tractor use in developing countries, *The Agricultural Engineer* (UK) **33**, 2, 52–4.

Inns, F. M. (1980). Animal power in agricultural production systems, *World Animal Review*, **34**, 2–10.

Kemp, D. C. (1980). Development of a new animal-drawn toolcarrier implement for dryland tillage. CEEMAT technical meeting II, Section 2.1, 6 March.

Okigbo, B. N. (1981). Alternatives to shifting cultivation, *Ceres*, November–December, 41–5.

Spoor, G., and Godwin, R. J. (1978). An experimental investigation into the deep loosening of soil by rigid tines, *J. Agric. Engng. Res*, **23**, 243–58.

Willcocks, T. J. (1981). Tillage of clod-forming sandy load soils in the semi-arid climate of Botswana, *Soil and Tillage Research*, **1**, 323–50.

2 Farmstead Tasks ✕

There are many farmstead tasks that it is necessary for the farmer to carry out, depending on the crop type, climate, and market requirements for his produce. Some of the more important tasks are considered in this chapter.

2.1 WATER SUPPLY

2.1.1 Domestic

The domestic water requirements of the farmer can be calculated from Table 2.1. If a hand pump is used for the water supply the flow a man can produce can be calculated from:

$$\text{Flow: l/s} = \frac{3.8}{\text{lift (m)}}$$

For example, for 20 metres lift a man can pump 0.2 l/s. (With this quantity he could irrigate 0.25 hectares dry-land crops in one day.)

TABLE 2.1. Domestic Water Requirements

	Quantity per day (litres)
Minimum water requirement for 1 adult for drinking, cooking and washing	9
Usual requirement including bathing: water carried	36–45
Usual requirement with plumbing	120–40
Bathtub (each time of use)	45–90
Flush toilet (each time of use)	14–18
Milk cow or ox	55–70
Horse	45
Pig	9
Sheep or goat	2

For an animal-powered system, the flow is given by:

$$\text{Flow: } l/s = \frac{30}{\text{lift (m)}}$$

Many other types of pumps are used, for example a ram pump using the energy in a flowing stream; water wheel-powered pumps; or plata type, wind-powered using a simple reciprocating pump. ITDG, among others, can supply plans for simple home construction.

2.1.2 Irrigation

In many parts of the world it is necessary to irrigate the crop if a reasonable yield is to be obtained. The principal types of irrigation systems in use are surface irrigation and overhead irrigation.

(i) Surface irrigation

(1) Basin irrigation. This takes a variety of forms:—types ranging from small depressions around trees, to a rice paddy—i.e. any ponded area or static water restrained by bunds.
(2) Border irrigation—free flow of water from outlets at the top of a slope.
(3) Furrow irrigation—water flow controlled, better than (2). The water inlet is commonly controlled by use of siphons of a suitable size to give the required flow rate.
(4) Trickle irrigation—very small flow rates directed to each plant. It is expensive in equipment but suitable for high-value crops.
(5) Subsoil water table control—a technique used in special circumstances.

These systems usually take water from a suitable river or canal and rely on gravitational flow. A small farmer would use these systems but would not usually have to provide a pump, although many large schemes have a large centralized pumping station.

(ii) Overhead irrigation. The water is pumped through nozzles in a jet or spray. Sprinkler irrigation has good application efficiency and is suitable for a wide range of crops, terrains, and soils. It does, however, have a high capital cost and high energy cost in running the pump. The systems can be used on small farms if the lower pressure systems are used. For small plots of 0.1 ha a single water-rotated sprinkler head will probably be sufficient if mounted 1–2 m above the crop on a 25 mm diameter standpipe.

When the land has received its water quota the sprinkler or the lateral line is moved to the next position arranged to give about a 50 per cent overlap on the wetted area. The space between the sprinklers on the laterals and the number of sprinklers on each depends on the required flow and pressure drop between the first and last to give reasonable uniformity of water distribution. A simple sprinkler layout is shown in Figure 2.1. The sprinkler system is designed to give a calculated depth of water at some application rate which depends on the soil type (see Table 2.2). The application efficiency is commonly 60–80 per cent; this is the proportion of water reaching the plant; the losses are mainly due to evaporation. The required depth of application depends on the soil characteristics and soil depth. These characteristics are:

(1) Field capacity—amount of water contained in the soil just before the drains start to flow.
(2) Wilting point—amount of water contained in the soil when plants just start to wilt.
(3) Permanent wilting point—plants cease to grow and will die.

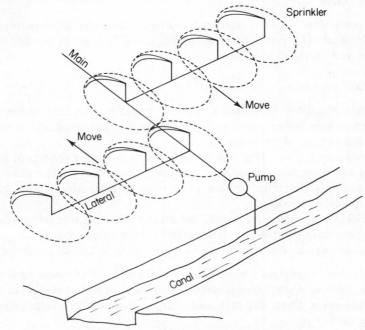

FIGURE 2.1 A simple sprinkler layout

TABLE 2.2 Water Intake Rates for
Overhead Irrigation

Soil type	Intake rate (mm/hr)
Clay	1–5
Clay loam	6–8
Silt loam	7–10
Sandy loam	8–12
Sand	10–25

Field capacity minus wilting point gives the available water measured in mm depth. If the available water on a day-to-day basis is less than that required by the plant, this must be made up by the irrigation water. Records over a number of years are necessary in order to plan the irrigation needs.

If the average depth of water required and the water intake rate of the soil is known then the length of time and the flow in litres per second can be calculated. For example, for 6.5 mm/hr soil intake rate and 80 mm required application depth, the time required is 12 hours. For an efficiency of 60 per cent the gross application rate will be 10.8 mm/hr. So for a 2 ha field the water supply will be

$$\frac{2 \times 10\,000 \times 10.8}{3\,600} = 60\,\text{l/sec.}$$

The crop water requirements depend on the transpiration rate, crop type, humidity, and wind speed, etc. The transpiration rate is measured in mm/hr water depth and can be calculated from information obtained from the rate of water evaporation from a pan of a particular design, such as American class A-type evaporation pans, and corrected by a factor called the crop coefficient. The crop coefficient is a factor used to convert the pan figure to a figure which would be obtained from the particular crop in question under the prevailing climatic conditions at the pan. From this information the irrigation requirement in mm/month can be calculated; in this example 80 mm requirement every 10 days will provide 240 mm of water per month. If the equipment can be moved twice a day the area that the line of sprinklers must cover will be one-twentieth of the area and the flow rate one-twentieth of the total or 3 l/sec.

From the manufacturer's information on sprinkler performance of flow and pressure, suitable sprinklers can be chosen. From the length of pipe, the pressure losses can be calculated and added to the pressure head giving the total pressure the pump must supply. The pump power, assuming 60 per cent efficiency, can then be calculated from

$$\text{kW} = \frac{P \times \text{Flow}}{1000 \times 0.6} \qquad \begin{array}{l} P \text{ in kN/m}^2 \\ \text{Flow l/sec} \end{array}$$

From the pump manufacturer's information, a pump can be selected and a suitable engine chosen to match its requirements of speed and power. (The head is the height of the water outlet above the pump, measured in metres). Table 2.3.

The engine used to drive the pump when running at reasonable efficiency will consume:

Diesel   250 gm per kW hr or 0.3 l/kWh
Petrol   330 gm per kW hr or 0.40 l/kWh

Therefore the fuel use per month can be calculated from the formula:

$$\text{Fuel/month} = P \times F \times H \times D$$

where P = Power in kW; H = Hours/day
F = 0.3 for diesel, 0.4 for petrol; D = Days/month.
The relationship between head and pressure is given by the formula:

$$\text{Head, m,} = 0.1022 \times \text{pressure, kN/m}^2$$
or
$$\text{Head ft} = 2.31 \times \text{pressure, lbf/in}^2$$

TABLE 2.3 Typical Pump Sizes and Powers

Power (kW)	Head (m)	Flow (l/min)	Pipe diameter (mm)
3.7	13	1065	75
	23	530	
1.5	16	400	50

(iii) Water Quantity. The amount of water available will have a major influence on the type of irrigation system that can be used. If the flow is small surface irrigation will be difficult although the effective flow can be increased by the use of storage ponds during periods when irrigation is not being practised.

Where water supply is limited it must be used with the highest possible efficiency. High efficiency is not usually obtained with surface irrigation unless design, operation, and management are of a high order and the feeder canals are lined to prevent seepage loss. Sprinklers and trickle irrigation generally give the highest efficiencies.

(iv) Water Quality. The presence of sediment in the water will cause problems with sprinklers and trickle systems; settling ponds and filters may be required. Very high levels of sediment can cause problems of silting up of canals and raising general land levels.

If the water contains objectionable material such as sewage or farm slurry then sprinklers are not recommended due to the smell and the depositing of unsuitable chemicals into the edible parts of the plants.

When salinity is a problem in the water or soil, or both, then surface application methods are generally preferred to perform a cheap leaching method. High concentration of salts in the soil or irrigation water causes problems due to toxicity, reduction in permeability, aeration, infiltration rate, and soil workability.

Many of these problems occur in arid and semi-arid areas where there is a high evaporation rate. The salt content of the soil is most conveniently determined by measuring the electrical conductivity of the soil water when the soil is at saturation level. The standard test is described in the USDA handbook (US Department of Agriculture).

The effects of salinity may be reduced to acceptable levels by arranging a careful balance between the amount of irrigation water supplied, the amount of water the plant uses and the amount of water entering the drainage system so that the salts are leached out of the soil and carried away in the drainage water, thus preventing excessive salt levels in the soil which will reduce the crop yields. The choice of crop is obviously important; a salt-tolerant crop should be grown in conditions where the salt levels are high. High salt levels can also cause problems of corrosion in the pumps, pipework, and sprinklers. In these cases the correct materials should be specified to reduce the effects.

An unpleasant side-effect of many irrigation schemes is an increase in waterborne diseases such as bilharzia and malaria. These effects can be minimized by the use of moluscicides and insecticides together with proper sanitation and washing arrangements for the population to reduce irrigation water pollution to an acceptable level.

In situations of high wind and high crops, sprinkler irrigation becomes less efficient and surface methods may be more appropriate.

2.2 SIZE REDUCTION OF HARVESTED PRODUCE

2.2.1 Plate mills

Figure 2.2 shows the principal components. The plates are made of cast iron, heat-treated, or hard-faced steel. Some larger type machines use artificial stone wheels made of various grades of corundum. The machines range in size from small hand-operated bench machines to larger engine-driven machines. The flour produced is rather coarse and sharp but it can be sieved and the coarse particles passed through again with a finer setting.

Typical power demands and outputs of plate mills:

3.7 kW   150 kg/hr to 250 kg/hr
10 kW   500 kg/hr air discharge

These mills can also be used with a wide plate setting for hulling maize.

2.2.2 Hammer mills

Figure 2.3 shows the principal components. The hammers rotate at 2000–5000 rpm depending on their rotating radius. The hammers are usually pivoted to reduce danger from foreign objects and are preferably hardened and balanced to give a long machine life. The feed inlet should be arranged so that the operator cannot put his hand in, and a stone trap and magnetic trap fitted. The

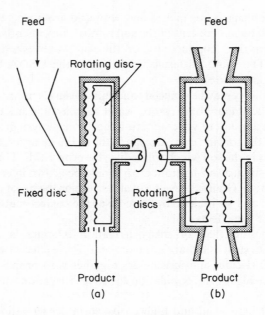

Feed   Feed

Rotating disc

Fixed disc   Rotating
                discs

Product   Product
  (a)   (b)

FIGURE 2.2 Plate mills

degree of fineness is controlled by the feed rate and size of holes in the retention screen, a low feed rate and small hole giving a fine product. Many hammer mills are fitted with a pneumatic extraction and meal-transport system using a cyclone at the bagging-off point. This adds to the power requirement and for a low-cost system can be dispensed with. A gravity feed system, both in and out, will give satisfactory results although it may be more dusty for the operator and good ventilation in the building will be essential. The flour produced can be much finer than with a plate mill but the power requirement for the same output is often three or four times greater. The minimum sizes are 2.5 kW with an output of 70 kg/hr. Some types of modern hammer mill do not have screens; this is quite an advantage as the screens are often subject to a lot of damage caused by stones and other rubbish.

A small livestock feed plant of 1 tonne/day capacity could consist of a 15 kW hammer mill and a 5 kW vertical auger mixer driven from one engine.

2.2.3 Roller mills

Figure 2.4 shows the principal components. This type is more usually found in the larger sizes. The larger the roller diameter the more effective the grinding action; the optimum speed ratio between the rollers is about 1.5–1. The sizes and angles of the corrugations are also important to the efficiency of the mill. Roller crushers are used on an agricultural scale for flaking grains, mostly for animal feed. Typically a 2 kW input gives 250–500 kg/hr output. One of the rolls is driven and rotates at

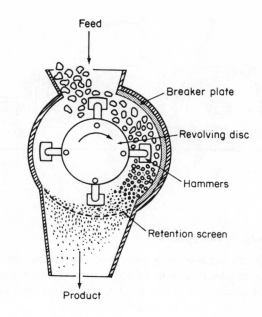

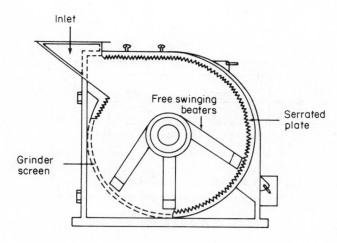

FIGURE 2.3 Hammer mills

400–450 rpm; the other is driven by frictional contact as the grain passes between the rollers. Roller shellers using hard rubber rolls rotating at different speeds, input speed 1500 rpm, are used to remove paddy husks; typical power is 2.5 kW for 1000 kg/hr.

Simple hand-operated roller crushers are sometimes used for cracking groundnuts to ease the task of husk removal.

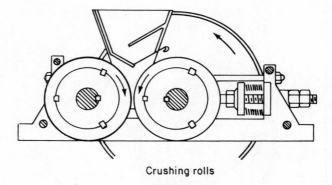

Crushing rolls

FIGURE 2.4 Roller mill

2.2.4 Rotating mortars and pestles

Figure 2.5 shows the principal components. Wood and stone construction is used for animal-powered machines, and metal for high-performance, engine-powered machines. This type of machine can be used for crushing copra and cider apples, and for grinding clays for brick-making.

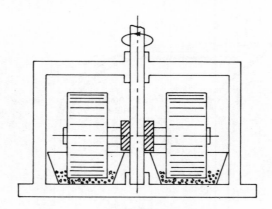

FIGURE 2.5 Rotary mortar and pestle

2.2.5 Rice hullers and millers

Figure 2.6a shows the principal components of the Engleberg type, using a solid rotor and adjustable knife; Figure 2.6b shows the principal components of a rubber-disc type which relies on impact for its effect.

The Engleberg type can be made very small, suitable for hand and pedal power, giving typically 5 kg/hr. Parboiling of the rice gives a better milling yield because the grain is tougher and sterilized so it stores better, is of improved nutritional value and has a firmer texture.

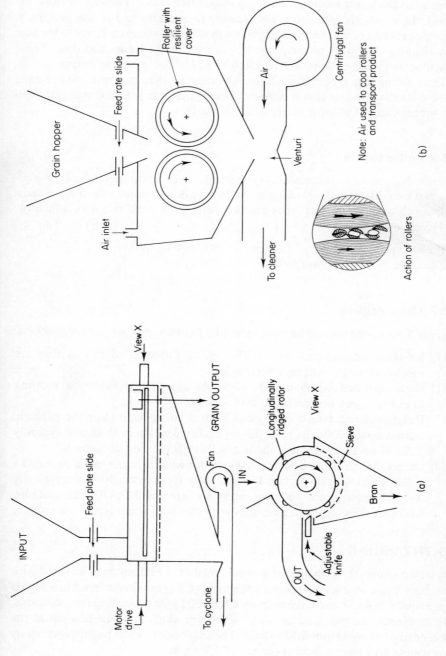

Grain hopper

Feed rate slide

Air inlet

Roller with resilient cover

Centrifugal fan

Air

Venturi

To cleaner

Note: Air used to cool rollers and transport product

Action of rollers

(b)

FIGURE 2.6b Rice huller with resilient roller

INPUT

Feed plate slide

View X

Motor drive

GRAIN OUTPUT

Fan

To cyclone

IN

Longitudinally ridged rotor

View X

Sieve

Bran

OUT

Adjustable knife

(a)

FIGURE 2.6a Rice huller with solid rotor and adjustable knife

The small motor-driven machines can typically produce 300 kg/hr for 10 kW input. The input shaft rotates at 700–800 rpm.

Some of the larger machines have a polisher attached and driven by the same motor. This consists of a drum with leather fingers attached to the periphery, rotating behind a screen, the action giving a final polish and extra lustre to the rice. Fan cleaning may also be incorporated to remove the husks and dust. These attachments increase the power required to 15 kW for the same output.

The rotating-disc type can cause greater damage if the conditions are not quite correct; its advantage is that it is much cheaper than the previous type although the rubber-faced disc must be replaced regularly.

2.2.6 Coffee hullers

These either consist of an embossed cylindrical roller working against a fixed edge, or a cast iron disc with bulbs cast into the face working against a fixed bar called a pulping chop. The simplest types can be operated by hand but with difficulty.

Typical outputs are:

Roller type	0.7 kW	120 rpm	500–600 kg/hr
Disc type	hand, wet feed		270 kg/hr

2.2.7 Maize shellers

Figures 2.7 a, b, and c show the principles of a number of types of maize shellers.

(1) Simple hand-operated sheller. The cobs are twisted slightly as they are pushed through: output 200 cobs/hr.
(2) Hand-operated bench model. The cobs are pushed against a rotating knobbly wheel: output up to 500 cobs/hr.
(3) Pedal-powered, twin-rotating disc with a fan to help clear the rubbish: output up to 6000 cobs/hr. If this type is fitted with a small engine, typically 1–3 kW up to 750 kg/hr cobs is possible under good conditions.
(4) Larger machines use a perforated cylinder with an auger inside or a spiky roller working against a check plate. These types of machines are typically 4–7 kW, input speed 600 rpm, with an output of 2500–3000 kg grain/hr. Many of these types of machines are fitted with a fan for air cleaning.

2.3 THRASHING

Figure 2.8 shows the principle of a small thrasher. The drum may be fitted with rasp bars, pegs, wire loops (specially for paddy). Engine-driven machines of this type require 2–4 kW and have outputs up to 250 kg/hr depending on conditions. The machines are mounted on small wheels or skids and can be used at the farmstead or moved from field to field. The small ones can be pedal-powered by two people and have a typical output of 150 kg/hr.

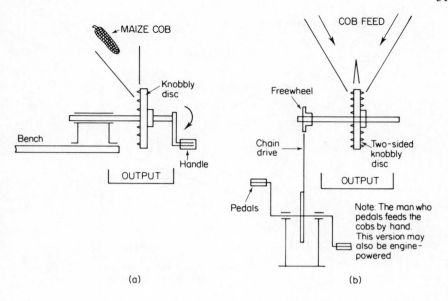

(a)

(b)

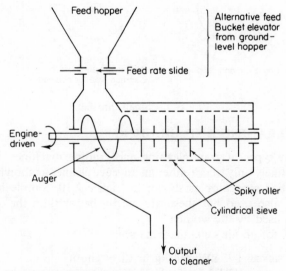

(c)

FIGURE 2.7 Maize shellers—(a) hand-powered; (b) pedal-powered; (c) engine-powered

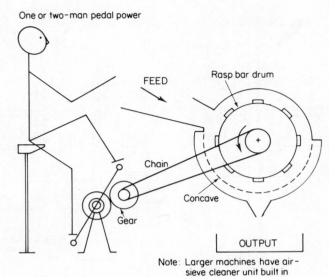

One or two-man pedal power

FEED

Rasp bar drum

Chain

Concave

Gear

OUTPUT

Note: Larger machines have air-sieve cleaner unit built in

Alternative drum and concave arrangements

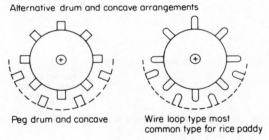

Peg drum and concave

Wire loop type most common type for rice paddy

FIGURE 2.8 Small thrasher

2.4 GRAIN CLEANING AND GRADING

There are many types of grain cleaning and grading machines. However, the small farmer is likely only to use either an air sieve cleaner as shown in Figure 2.9 or a fluidized bed separator as shown in Figure 2.10. Simple hand-powered winnowers are also used but these are hardly better than the simple, round shallow basket used by the women.

Typical powers for an air sieve machine are:

   1 blower, 2 sieves, 1–3 kW, 600 kg/hr throughout
   Rotary screen type (IRRI Design), 2 kW, up to 2.5 tonnes/hr.

2.5 BUSH CLEARING

Brush cutters and saws: Figure 2.11a shows a typical chain saw and Figure 2.11b shows a brush cutter. They are powered by small two-stroke petrol engines, of

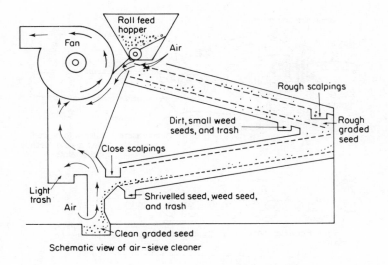

Schematic view of air-sieve cleaner

FIGURE 2.9 Air-sieve cleaner

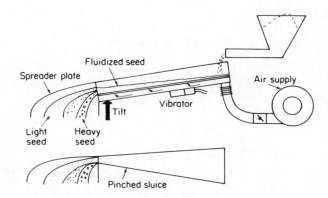

FIGURE 2.10 Fluidized bed separator

30–50 cc, and have various attachments such as saw blades, grass-cutting blades, and chain saws.

2.6 OTHER TASKS

Some other machines which may be required by the farmer are:

    hole diggers;
    concrete mixers;
    electrical generators;
    kenaf ribboners;
    powder dressers;

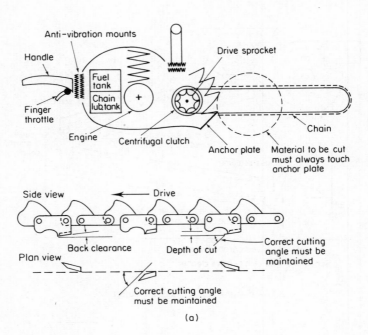

FIGURE 2.11a Chain saw

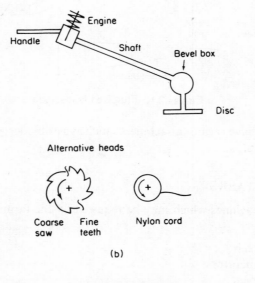

FIGURE 2.11b Brush cutter

pelleting press;
root choppers;
paddy separator zig-zag table;
root peelers;
cylindrical rasp graters for cassava;
groundnut decorticator—reciprocating bar and semi-circular sieve system;
cane crushers.

REFERENCES

Tools for Progress, Intermediate Technology Development Group Ltd., 9 King Street, London, WC2E 8HN.
US Department of Agriculture, *Diagnosis and Improvement of Saline and Alkaline Soils*, Agricultural Handbook No. 60.
Withers, B., and Vipond, S. (1980). *Irrigation Design and Practice* (Batsford).

3 *Transport Tasks and Conditions*

Effective rural transport requires an infrastructure of all-weather roads

3.1 INTRODUCTION

Transportation is a vital part of many production systems, as products which remain at the site of production cannot normally be regarded as having entered the marketing system. The initial change of location from production site to the first stage of the marketing infrastructure is therefore a significant one.

Transportation from the smallholder sector of agriculture in developing countries is a particularly important example. In view of the need to increase the supply of basic foodstuffs from within the country in order to support growing urban populations or for revenue-earning exports, it is widely recognized that the small-scale farmer should be encouraged to improve present (typically) low yields and grow more produce than is required for subsistence at the house-hold/village level.

The excess produce has virtually no value, however, unless it can be introduced into the marketing system which, in many cases, involves a relatively short journey to a small town market or to a main road pickup point. This can be thought of as the primary transportation phase.

The later transportation phases are usually well catered for in most developing countries, being carried out by buses, lorries, or pickup trucks. However, the primary phase usually involves the carriage of small, irregularly phased loads over very poor roads or tracks. These characteristics are far from the ideal requirements of, for example, a transport contractor with a pickup truck who seeks regular capacity loads to be carried at high speeds on good surfaces.

As a result there is often little motivation for the smallholder to produce excess outputs since there is no guarantee that transport facilities will be forthcoming when required or, if they are, that the rates charged will enable a profit to be made on the produce when sold. The most vital phase of the transport system is therefore often the most difficult.

While arterial roads in many developing countries are excellent and main roads usually well maintained for the major part of the year, local roads at farm and village level are naturally of low priority for upgrading and maintenance. It is sensible to accept that the physical characteristics of these roads, apart from the improvement of water crossings and other obvious constraints, will tend to remain substantially the same for the forseeable future, and that attention should therefore be directed to the design and selection of vehicles suitable for operation on these roads for all or a major part of the year.

3.2 THE AGRICULTURAL ENVIRONMENT

A review of the characteristics of smallholder agriculture in developing countries has been given in Chapter 1, from which it may be deduced that the factors tending to affect the transportation requirements for agricultural produce and inputs are:

(1) Level of technology applied.
(2) Farm or holding areas.
(3) Geographical position of holdings relative to the homestead.
(4) Geographical position of homestead relative to the village and to the local market/main road.

On farm transport at and above subsistence level is associated with the provision of crop inputs, the collection of harvested produce, and the supply of household requisites such as firewood and water. Loads will tend to be small (10–100 kg) and distances short (1–3 km), normally in off-road conditions. Such requirements can usually be satisfied by non-vehicular transport modes apart from the case of intensive cash-crop farming significantly above the subsistence level.

Off farm transport above the subsistence level ranges from occasional trips to local markets for household and farm requisites, up to regular and considerable transport to market or collection centres with cash-crop loads such as cotton, coffee, tea, maize, etc. together with return transport of fertilizers, chemicals, seeds, and other inputs.

Loads will, however, still tend to be fairly small (50–200 kg) and distances fairly short (5–25 km) partly on-road and partly off-road (IBRD, 1977). The provision of economic and reliable transportation services to cater for such conditions is a difficult problem, the solution of which can be aided by information about likely vehicle operating costs.

3.3 ECONOMIC JUSTIFICATION OF A VEHICLE

The justification for using a vehicle to transport a load from, for example, a farm or village to a local market can be assessed either on an economic or a financial basis. The increase in value of the load due to its change of location must be greater than the cost of its transportation. On an economic basis this is calculated in relation to the value to the nation of increased domestic food supply (import substitution) and the costs to the nation of the transport operation (imported fuel, vehicles, etc.). Decisions may therefore be taken at governmental level to control transport facilities or commodity prices, for example, or to subsidize transport fuel costs.

However, in a free-enterprise situation, which can often be a suitable basis for small-scale transportation systems, it is the financial aspect which is likely to dictate success or failure. Where the transport fee charged allows the farmer to make a profit on the sale of the product and also allows the contractor to cover his operating costs and make a profit, the financial basis for a satisfactory service exists; the potential results are more produce available on the domestic market and more profit realized in the smallholder sector—both desirable features. Information on load values and probable operating costs, however, is of great importance in assisting the decisions on vehicle purchase and operation (Crossley, Kilgour, and Morris, 1976; Morris, 1975; Crossley, 1978).

3.4 DESIGN AND SELECTION OF VEHICLES

In the preceding section it was concluded that a transport operation is financially justifiable when the increase in the value of the load due to its change of location is greater than the cost of transportation (see Figure 3.1).

There are therefore two aspects to a decision about vehicle use. The farmer or load-owner will need to assess the increase in value of his load due to its being transported. For example, a crop may be worth $10 per tonne in the village but $15 per tonne in the town 5 kilometres away. The increase in value is thus $5 per tonne or $1 per tonne kilometre. If the fee charged for transportation is less than this then the farmer will consider using the transportation service.

Meanwhile, the vehicle owner needs to be able to calculate the cost per tonne kilometre of running his vehicle so as to ensure the fee he charges is somewhat higher than the costs, allowing an operating profit to be made.

The cost of transportation is conventionally expressed in currency units per unit of payload weight and unit of distance, for example in US$ per tonne kilometre ($/tonne km), and may be found by dividing the total costs (fixed plus

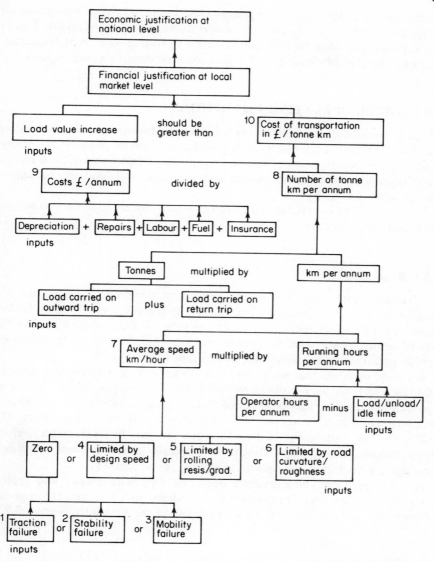

FIGURE 3.1 Economic/financial justification of a transport vehicle

variable) incurred during a unit time (e.g. one year or one year quarter) by the product of load carried and distance covered during the same unit time, as illustrated in Figure 3.1 (blocks 8, 9).

The distance covered, however, depends on the running time and the average speed, which in turn is affected by the characteristics of the vehicle and the road conditions (gradient, surface, horizontal curvature, roughness, etc.) in which it operates. (In certain conditions the vehicle may be unable to perform at all due for example to inability to negotiate obstacles, or due to traction limitations on upslopes in slippery conditions, or to lateral stability limitations on sideslopes.)

It is therefore evident that, to enable vehicle operating costs to be calculated, data on a number of physical and socio-economic conditions are required.

3.5 VEHICLE OPERATING CONDITIONS

Although smallholder agriculture occurs in many developing countries in many parts of the world, each of which has a unique blend of physical and socio-economic conditions, it is possible to identify those factors which will have a direct bearing on vehicle performance and operating costs (and also to suggest typical values for each).

The factors may be separated into three classes, namely road data, time/cost data, and load/trip data.

3.5.1 Road data

The factors of importance are the average gradient of the road, the curvature, the roughness, the rolling resistance and the tyre grip conditions. All these factors will affect the average speeds of vehicles over the road and some will affect the cost of fuel and spare parts. Ways of determining these factors and the effects on vehicle performance are described in the Appendix.

It should be noted that some road conditions (rolling resistance, roughness, and tyre grip) will vary with the seasonal climatic changes. In the absence of specific information it may be assumed that a typical climatic pattern consists of one very wet quarter year, one fairly wet quarter and two dry quarters. The rolling resistance categories in these four quarters may be taken respectively as 'high', 'medium', 'low', 'low'. Roughness is typically 'high', 'medium', 'medium', 'medium'. Tyre grip is 'very poor', 'medium', 'good', 'good'.

Average gradient and curvature does not vary with season and typical category values are 'low' for gradient and 'medium' for curvature. Descriptions of the road condition categories may be found in the Appendix, Section 3.7.9.

3.5.2 Time/cost data

The number of hours worked by the vehicle operator, the wages paid, fuel cost, and the cost of spares and repairs will affect the overall cost of operation of vehicles. The Appendix (Sections 3.7.7 and 8) details the effects.

Typical values for hours worked (including loading, unloading, and waiting time) and operator wages are 2400 hours per year at $0.5 per hour. Fuel costs vary widely, depending on whether or not the country produces oil. Typically, however, diesel fuel will tend to be subsidized for agriculture and transport, whereas petrol will not.

3.5.3 Load/trip data

The average distance of the round trip from village to market, together with the time spent in loading, unloading, and waiting during each trip will affect the number of trips per day which a vehicle could possibly make, while the loads available on the outward and return journeys will determine how many trips the vehicle will *actually* make.

The usual trip distance is found to be between 5 and 10 kilometres and the usual load found to be available per day from 5–10 farms in all seasons is in the range 0–2 tonnes. The typical time spent per trip in loading, unloading, and waiting can be taken as between 1 and 2 hours.

3.6 A TECHNIQUE FOR THE CALCULATION OF VEHICLE OPERATING COSTS IN SPECIFIED CONDITIONS

The performance and operating cost characteristics of various vehicles will depend on the operating conditions. The factors described in the previous sections can be incorporated into a useful general technique for predicting operating costs.

3.6.1 Principle

Figure 3.1 provides a basis for calculating the final cost (in currency units per tonne kilometre) of moving a load with a given vehicle in specified conditions. By referring to the information in the Appendix detailed data may be derived. The general principles, however, may be stated as follows.

First, estimate the average speed at which the vehicle will travel by comparing the design speed, the rolling resistance/gradient speed, and the road curvature/roughness speed (blocks 4, 5, 6 in Figure 3.1), and selecting the lowest speed. This is used as the average speed (block 7, Figure 3.1). Note that additional accuracy may be achieved if required by performing this calculation over four year quarters, uphill and downhill, for outward and return trips.

Second, divide the round trip distance by the average speed (from previous section) to get the number of hours per trip. Add the loading, unloading, and idle time per trip to get the total time per trip. Divide this into the operator hours per day, which gives the number of possible trips per day.

Third, check whether the vehicle will run out of load by multiplying the possible trips per day by the vehicle payload and compare the result with the available load per day. If the available load is the larger, select 'possible trips per day' to be actual trips per day. If the available load is smaller then divide it by vehicle payload and use the result as actual trips per day. (This need not be a whole number, as the trips will average out over a period.)

Fourth, multiply the actual trips per day by the round trip distance to give the number of kilometres run per day. By adding together the loads carried on the actual and return trips and multiplying by the number of kilometres run it is possible to calculate the number of tonne kilometres run per day. Multiplying this by the number of days worked per year gives the number of tonne kilometres per year.

Finally, the total operating cost of the vehicle in a year is obtained by adding depreciation (i.e. purchase price divided by years of life), repairs (expressed as a proportion of depreciation), labour (i.e. wages per hour multiplied by number of hours worked), fuel used and other factors such as insurance. This gives the cost per year (block 9, Figure 3.1) and, when this is divided by the number of tonne kilometres per year (from the previous section), the final result of cost per tonne kilometers (block 10) is obtained.

3.6.2 Example

An example using the same section system as above will help to illustrate this technique.

First, the vehicle design speed is given as say 10 km/h, the speed imposed by the rolling resistance and gradient effects is 7 km/h, and the speed to which a vehicle would be restricted by road roughness and/or curvature is found to be (say) 30 km/h. Selecting the lowest gives a result of 7 km/h for block 7.

Second, if the round trip distance is 10 km the running hours per trip is $10 \div 7$, that is 1.43 hours. Adding (say) 1 hour for loading, unloading and waiting time gives a trip time of 2.43 hours. If the operator works for 10 hours per day (including loading, unloading, and waiting time) then the number of possible trips per day is $10 \div 2.43$, that is 4.1.

Third, possible trips per day (4.1) multiplied by vehicle payload (say 0.4 tonnes) gives 1.64 tonnes per day. If the available load per day was only 1 tonne then the *actual* trips per day would be $1 \div 0.4$, that is 2.5. In this example, however, we will take the available load as 5 tonnes, thus the *actual* trips per day remain at 4.1.

Fourth, the distance run per day is therefore 4.1×10, that is 41 kilometres per day. If the vehicle is full (0.4 tonnes) on the outward trip and half full (0.2 tonnes) when returning, the number of tonne kilometre per day is $(0.4 + 0.2) \times 41$, that is 24.6 and working (say) 280 days a year gives a load total of 24.6×280, that is 6888 tonne kilometres per year.

Finally, if the vehicle costs $5000 and lasts for five years the depreciation is $1000 per year. Repairs will be taken as $1500 per year, labour (say $0.5 per hour, 10 hours per day, 280 days per year) is $1400 per year, fuel used (say 1.5 litres per hour, 1.43×4.1 hours per day, 280 days per year) is 2460 litres or (at $0.17 per litre) $1720 and insurance etc is (say) $250 per year. The total cost in a year is therefore $5870. The number of tonne kilometres per year was found in Section 4 to be 6888. The final cost per tonne kilometre is therefore 5870/6888, that is $0.85/tonne km.

3.6.3 Application of the technique

The above exercise can be carried out for any number of alternative vehicles, thus providing a way of assessing which alternative is the cheapest per tonne kilometre in the specified conditions. An example of the application of a similar (computer-aided) technique to predict the performance and costs of three vehicles in 'typical' specified conditions appears below. (The Appendix gives more detailed information about the condition factors described in the earlier sections of the present chapter.)

The three vehicles selected for the example are a small single-axle tractor trailer combination (representing the lower end of the smallholder transport sector), a 1 tonne pickup truck (representing the upper end), and a specialist vehicle design proposal code-named SERPENT (representing an attempt by the author to produce a vehicle specification to suit smallholder conditions). Details of the vehicles are given in Figure 3.2 and Table 3.1.

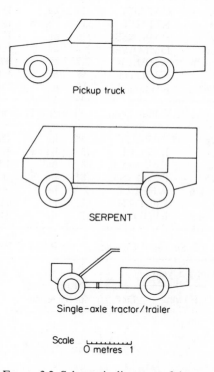

Pickup truck

SERPENT

Single-axle tractor/trailer

Scale
0 metres 1

FIGURE 3.2 Schematic diagrams of three
vehicles

The conditions data used were obtained through a survey of postgraduate students from developing countries at NCAE, Silsoe. Respondents were asked to divide the (seasonal) year into four quarters and to specify how many quarters were typically very wet, fairly wet, and dry respectively. Ignoring the order of occurrence, the possible combinations total fifteen. It was found however that only nine combinations appeared in the returns and, of these, five were most significant (Table 3.2). The pattern of one very wet quarter, one fairly wet and

TABLE 3.1 Data Files for Three Vehicles

Data File Values Used for Single-axle Tractor/Trailer

WEIGHT	250	kg	CAPACY	300	kg	IFRONT	2	
FRODIA	.574	m	RTFUEL	.42	l/kwh	IREAR	0	
READIA	.65	m	HMMASS	.65	m	FDRIVE	2.	
POWER	5.2	kW	DMMASS	1.2	m	RDRIVE	0.	
SPEED	10.	km/h	WHEELB	1.9	m	TRANEF	.9	
CVEH	1000.	£	FWIDTH	.2	m	HOMLOD	.8	m
YLIFE	5.	Years	RWIDTH	.2	m	DOMLOD	.05	m

Date File Values Used for Pickup Truck

WEIGHT	1090	kg	CAPACY	1000	kg	IFRONT	0	
FRODIA	.65	m	RTFUEL	.38	l/kwh	IREAR	2	
READIA	.65	m	HMMASS	.76	m	FDRIVE	0.	
POWER	46	kW	DMMASS	1.4	m	RDRIVE	2.	
SPEED	100.	km/h	WHEELB	2.78	m	TRANEF	.9	
CVEN	4000.	£	FWIDTH	.19	m	HOMLOD	1.1	
YLIFE	10.	Years	RWIDTH	.19	m	DOMLOD	.07	

Data File Values Used for SERPENT Vehicle

WEIGHT	700.	kg	CAPACY	700.	kg	IFRONT	0	
FRODIA	.72	m	RTFUEL	.3	l/kwh	IREAR	2	
READIA	.72	m	HMMASS	.8	m	FDRIVE	0.	
POWER	7.8	kW	DMMASS	.85	m	RDRIVE	2.	
SPEED	26.	km/h	WHEELB	2.3	m	TRANEF	.9	
CVEN	2500	£	FWIDTH	.2	m	HOMLOD	.9	m
YLIFE	10.	Years	RWIDTH	.2	m	DOMLOD	.85	m

TABLE 3.2 Seasonal Characteristics Rainfall Patterns

Quarters per year			Number of occurrences	Total %
V. wet	F. wet	Dry		
1	1	2	31	49
1	2	1	7	11
0	2	2	7	11
2	1	1	5	8
1	0	3	5	8
2	0	2	3	5
0	1	3	3	5
Others			2	3

two dry occurred as much as all the other combinations added together and may therefore be taken as a typical pattern.

Typical smallholder transport conditions take place over roads with low-to-medium gradients and with medium curvature. In the very wet quarters rolling resistance is high, roughness is medium to high, and tyre grip is very poor.

The fairly wet season shows an improvement to medium for all parameters while in the dry season rolling resistance is low and grip is good; roughness, however, remains at a medium level.

The usual return-trip distance is found to be between 5 and 10 kilometres and the usual load said to be available per day from 5–10 farms in all seasons is in the range 0–2 tonnes.

The typical time spent in loading, unloading, and waiting is given as between 1 and 2 hours (times over 8 hours ignored), while the operator working hours per quarter tends to peak at around 600 (equivalent to 10 hours per day, 5 days per week, 12 weeks per quarter). This includes loading/unloading and waiting time.

The stated operator wages per hour ranged up to £1.80 but the majority will be paid less than £0.5 with the highest number getting less than £0.2 per hour.

Petrol prices in some (oil-producing) countries are stated to be less than £0.1/litre (£0.45 per gallon) while in a few others the price rises to £0.6/litre. In most countries, however, petrol consumers (i.e. private motorists and light goods vehicle operators) pay between £0.2 and £0.3/litre (£0.9 to £1.36 per gallon).

Diesel fuel prices, on the other hand, exhibit the effect (presumably) of government subsidy of the commercial transport and agricultural sectors, indicating that most consumers pay less than £0.3/litre and the most usual price is around £0.1/litre (£0.45 per gallon). (All prices are based on late 1979–early 1980 experiences.)

This kind of factor points to the effect that government economic measures, exerted through financial means, can have on activities such as transportation and indicates that care must be taken not to recommend certain solutions (e.g. vehicles) solely on the basis of artificially favourable factors (such as subsidized diesel fuel) which may be changed without warning. For this reason, all computations of alternative vehicle costs described in this example assume that petrol and diesel fuel prices are equal.

The data files describing the three vehicles mentioned earlier (namely the single-axle tractor/trailer, the pickup truck, and the SERPENT) were used in conjunction with the 'typical' conditions mentioned above (these can be expressed in a conditions data file as in Table 3.3). The results show that the costs per tonne kilometre for the three vehicles (Figure 3.3) all reduce in a similar way as road conditions improve from the wet quarter through the fairly wet and into the two dry quarters. The cost per tonne kilometre for the single-axle tractor combination, however, is substantially higher than those for the other two vehicles, due mainly to its low speed and small payload.

It may be seen that if the pickup were allowed to serve two villages rather than one, which its less than 50 per cent utilization (mentioned later) would allow, its costs would be lower than that of the SERPENT. For one village as specified, however, the SERPENT incurs rather lower costs.

A display of machine utilization (Figure 3.4) shows that the available load occupies the SERPENT for between 60 and 70 per cent of its time, while the pickup is used for less than 50 per cent. The single-axle tractor, however, is occupied 100 per cent of its operator's specified time and, as illustrated in Figure

TABLE 3.3 Data File Values Used for 'Typical' Conditions

Identifier and Units		Quarter 1	Quarter 2	Quarter 3	Quarter 4
GRADE	degrees	.5	.5	.5	.5
CURVE	km/h	75	75	75	75
RRFAC		1.6	.7	.15	.15
RREXP		.7	.55	.3	.3
ROUGH	km/h	65	75	75	75
OUTLOD	kg	1000	1000	1000	1000
RETLOD	kg	500	500	500	500
HORDAY	h	10	10	10	10
DAYWEK	days	5	5	5	5
WEKQTR	weeks	12	12	12	12
CWAGES	£/h	.3	.3	.3	.3
DISOUT	km	5	5	5	5
DISRET	km	5	5	5	5
HORLOD	h	1.5	1.5	1.5	1.5
HORIDE	h	1	1	1	1
FUELCO	£/l	.25	.25	.25	.25
COHES	kN/m²	5	20	20	20
TANPHI		.6	.4	.4	.4
DEFCON	m	.04	.02	.02	.02
AREFAC		1.22	1.00	1.00	1.00
FACSUR		.5	.75	.75	.75
SERFAC		3	2	2	2

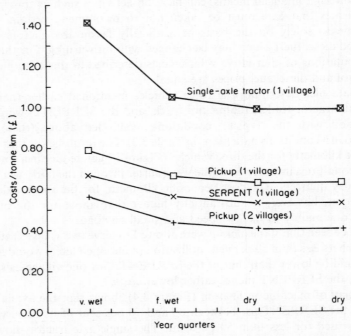

FIGURE 3.3 Costs per tonne km for three vehicles in 'typical' conditions

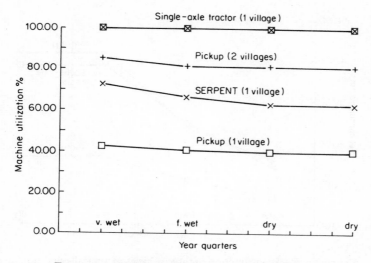

FIGURE 3.4 Machine utilization in 'typical' conditions

3.5, is only able to shift about 50 per cent of the load available. To satisfy the demand, therefore, two of these vehicles would be required.

Figure 3.6 shows that the fuel consumptions in miles per gallon of the SERPENT and of the single-axle tractor are both good compared with that of the pickup truck but, as illustrated in Figure 3.7 because two single-axle tractors are required the total fuel consumed in moving all the available load is higher than for the other two vehicles, which contributes to the very high cost incurred in shifting the load. The SERPENT and pickup truck are seen to incur similar costs, with the SERPENT ahead on points.

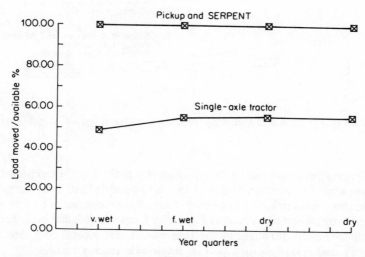

FIGURE 3.5 Proportion of available load shifted in 'typical' conditions

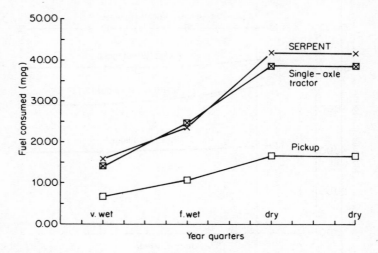

FIGURE 3.6 Fuel consumption of three vehicles in 'typical' conditions

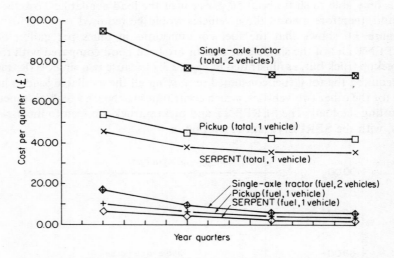

FIGURE 3.7 Total and fuel costs for three vehicles in 'typical' conditions

The conclusion which can be drawn from this preliminary exercise is that the important cost is that incurred in shifting the available load over the necessary distance; this cost is reflected in the cost/tonne km results but is best brought to prominence by illustrations such as Figure 3.7 from which it can be deduced that to shift 1 tonne/day throughout the year would cost a total of £1560 using a SERPENT and £3190 using a pair of single-axle tractor/trailers.

APPENDIX: 3.7 DERIVATION OF PERFORMANCE AND COST FACTORS

As illustrated in Figure 3.1 and discussed in Section 3.6, predictions of vehicle operating costs can be produced by manipulating data provided from the interrelation of a considerable number of technical and socio-economic factors. These factors are described more fully in the following sections.

3.7.1 Gradient measurement/estimation

The average gradient of a road may variously be expressed as an angle (in degrees), in terms of the ratio of rise to unit length (e.g. 1 in 4), as a percentage, or as a direct statement of rise per unit length usually expressed in metres per kilometre (it is assumed that the horizontal road length is used throughout, i.e. the tangent of the angle rather than the sine).
Some comparisons are shown in Table 3.4.

TABLE 3.4 Methods of Expressing Gradient

Angular (degrees)	Ratio	Percentage (%)	Rise (m/km)
2	1 in 28.65	3.49	34.92
5	1 in 11.07	8.75	87.49
10	1 in 5.76	17.63	176.32
15	1 in 3.73	26.79	267.95

Three methods of gradient measurement are possible. One way is to measure the actual road profile on the ground using a surveying level or theodolite. A second method is to examine aerial photographs with the aid of stereoscopic devices. The third method, which is simple and fairly accurate, is to use 1:50000 scale maps (or larger scale if available) of the type produced by the department of surveys in many developing countries. Such maps are based on aerial photographic surveys and display contour lines, typically at 50 metre intervals.

By noting where contour lines cross a selected road it is possible to measure and accumulate the total rise and the total fall along the road length. Dividing this by the road length gives the average gradient expressed in metres per kilometre. Improved accuracy can be achieved by interpolating between cutting and non-cutting contour lines and by the use of spot heights where available.

When the values of rise and fall have been found they can be converted to an angular gradient by the formula

$$\beta = \tan^{-1}\left(\frac{R+F}{2}\right)$$

where β is average gradient in degrees
R is total rise in metres per kilometre
F is total fall in metres per kilometre

Where measurements cannot be taken, a road may be classified into one of four categories as shown in Table 3.5. For descriptions of the four categories, refer to Section 3.7.9.

TABLE 3.5 Gradient Categories

Gradient category	Average gradient (°)
Low	0.5
Medium	1.5
High	3.0
Very high	5.0

3.7.2 Effect of gradient on a vehicle

When ascending an upgrade a vehicle experiences two effects. The first is an increase in resistance due to the downslope effect of its weight (i.e. $G1 \sin \beta$) and the second is a transfer of some weight from the front to the rear wheels, amounting to

$$G1 \cos \beta \, (1 - C1/W) + G1 \, H1/W \, \sin \beta - G1 \, (1 - C1/W)$$

where $G1$ is the weight of vehicle plus payload acting at a height $H1$ and a distance $C1$ from the rear axle, W is the vehicle wheelbase and β is the slope angle.

Where a trailer is being pulled its downslope effect ($G2 \sin \beta$) makes itself felt on the vehicle both as an additional resistance or drawbar load and as an additional weight transfer on to the vehicle rear wheels due to the drawbar load.

3.7.3 Rolling resistance measurement/estimation

Resistance to the forward motion of a tyre when rolling on a road surface occurs in two ways—as a result of the hysteresis loss in the tyre walls due to flexing, and as a result of deformation of the surface (such as soil) over which the tyre is passing. In off-highway situations the soil deformation loss is very much higher than the internal tyre losses (Inns and Kilgour, 1978) while on hard surfaces the tyre energy losses are greater.

Direct measurement by towing the vehicle may be performed where possible. Rolling resistance tends to increase in direct proportion to wheel loading, and the approximately constant ratio of rolling resistance to wheel loading is denoted rolling resistance coefficient ψ.

In the absence of field information, estimates of this coefficient may be used to predict the rolling resistance of a given tyre size in given conditions, see Figure 6.23.

3.7.4 Vehicle speed as restricted by gradient and rolling resistance

If weight transfer effects are ignored (acceptable for small gradients) the 'gradient resistance' of a vehicle simplifies to $G1 \sin \beta$ where $G1$ is the vehicle weight in kgf.

This result should be converted to kN by multiplying by 0.0098 (on downgrades the gradient effect is taken as negative).

The rolling resistance of the vehicle may be derived by measurement or by estimation of the rolling resistance coefficient ψ from Figure 6.23 which is then multiplied by the vehicle weight $G1$. (Unequal size wheels should be calculated individually.)

This result should also be converted to kN. The sum of gradient resistance and rolling resistance is then divided into the vehicle power (kW) to find the maximum speed at which the vehicle could ascend the gradient. The result is in metres per second, which can be converted to km/h by multiplying by 3.6.

For example, if a vehicle of mass 3 tonnes is ascending a 3° gradient on a surface with a rolling resistance coefficient ψ of 0.25 and the engine power delivered to the wheels is a maximum of 40 kW, at what sustained speed could the vehicle ascend the gradient?

The 'gradient resistance' GL Sin β is $3000 \times$ Sin 3°, that is 157 kgf or 1.54 kN. The rolling resistance is 0.25×3000, that is 750 kgf or 7.36 kN. Thus the total resistance is 8.9 kN and this, divided into the engine power, gives a speed of 4.5 m/s or 16.18 km/h. (Note that on a downgrade the total resistance is 7.36–1.54, that is 5.82 kN and the speed attainable is $40 \div 5.82 = 6.87$ m/s or 24.7 km/h.)

3.7.5 Road curvature effects

Again three methods, using direct measurement, aerial photographs, or maps, are possible. Direct measurement on the ground can be carried out with the use of a prismatic compass or a theodolite. Indirect measurements are best performed using maps (since aerial photographs usually offer no advantages of accuracy for curvature measurements), and may use one of several techniques. Two main types exist—that developed for assessing meandering rivers (where a detour index or sinuosity index is expressed as the ratio of actual distance to direct distance), and that used for road measurement (Robinson, Hide, Hodges, Rolt and Abaynayaka 1975; Hide, Abaynayaka, Sayer and Wyatt, 1975) in which the road is divided into a series of circular arcs and the subtended angles are measured by protractor. The cumulative angle in degrees is divided by the road length to give an average curvature value in degrees per kilometre.

Again, in the absence of specific data from measurements it may be necessary to place roads into curvature categories and to select interim values for curvature speeds, Table 3.6. (Description of the categories appears in Section 3.7.9.)

3.7.6 Roughness effects

Road roughness may be continuous, as in the case of textural characteristics of the construction material, or intermittent, as with potholes. A number of methods have been used to measure surface roughness, and the Transport and Road Research Laboratory have used both towed and mounted bump integrators in conjunction with moving vehicles. Surface roughness is then expressed in

TABLE 3.6 Vehicle Speed and Road Curvature

Curvature (°/km)	Curvature category	Vehicle speed limitation (km/h)
0	Low	80
100	Medium	68
200	High	58
300	Very high	45

millimetres per kilometre, ranging from 1800 mm/km in the case of an asphaltic concrete road to 10000 in the case of a gravel road in poor condition.

Speed restrictions due to roughness may again be categorized, Table 3.7.

Vehicle design speed. Due to the overall gearing from a limited range of ratios many vehicles, particularly small two-wheel and four-wheel tractors, are limited to a maximum 'design speed'. This is the speed which may be achieved unladen on a level road at maximum engine speed in top gear.

3.7.7 Fuel costs

The rate of consumption of fuel depends on the amount of power being used and the type of engine. The rate of fuel used per unit of power ranges from 0.5 l/kWh for a small petrol engine to 0.29 l/kWh for an automotive diesel engine. Values of rate of fuel which can be used for different engines are shown in Table 3.8. The rate of use (specific fuel consumption) will also usually increase at less than full

TABLE 3.7 Vehicle Speeds and Road Roughness

Roughness (mm/km)	Roughness category	Vehicle speed limitation (km/h)
2,000	Low	80
5,000	Medium	68
10,000	High	58
15,000	Very high	45

TABLE 3.8 Rate of Use for Fuel per Unit of Power for Various Engine Types

Type of engine	Rate of fuel use at full power (l/kWh)
Automotive diesel	0.29
Medium diesel (5–20 kW)	0.32
Small diesel (< 5 kW)	0.35
Automotive petrol	0.38
Small petrol (< 10 kW)	0.50

power and an expression was derived for fuel use in litres per hour, given by

$$\text{Fuel used} = \text{Rate of fuel use} \times \text{Power used} \times \left(\frac{\text{Power used}}{\text{Full power}}\right)^{-0.15}$$

In the absence of accurate information about fuel consumption rates an approximation (average operation) can be achieved by selecting the figure of l/kWh for the specified engine from Table 3.8 and multiplying by engine power $\times 0.6 \div (1-s)$ where s is wheel slip value. Approximate values of slip s for three types of vehicle in four road grip conditions are given in Table 3.9.

TABLE 3.9 Vehicle Slip Values and Road Conditions

Vehicle type	'Good' grip conditions	'Medium' grip conditions	'Poor' grip conditions	'Very poor' grip conditions
Single-axle tractor/trailer	0.05	0.10	0.20	0.40
Pickup truck	0.02	0.08	0.15	0.25
Large tractor/trailer	0.01	0.05	0.10	0.20

For example, a 40 kW pickup truck in medium grip conditions would use 0.38 l/kWh (automotive petrol engine) at full power (Table 3.9). The value for slip s (Table 3.9) would be 0.08. Thus the fuel level consumed would be

$$0.38 \times 40 \times 0.64 \div (1-0.08)$$

that is

$$9.9 \, l/h$$

More accurate methods for determining values for wheel slip may be found in Chapter 6.

3.7.8 Depreciation and spares/repairs costs

A simplified method of calculating depreciation is on a straight line basis over the estimated life of the vehicle. For example the depreciation on a $5000 vehicle lasting 8 years would be $5000 \div 8$, that is $625 per year.

To this should be added 25 per cent for insurance, etc., making $780 per year. Spares and repairs costs may be specified as a proportion of depreciation. A spares/repairs factor of (say) 1.5 would give a cost per year in the above example of 1.5×625, that is $938 per year.

In practice the spares/repairs factor selected will depend considerably on the roughness of the road conditions and partly on the amount of use of the vehicle. Assuming fairly extensive operation (1500 hours per year) the factors may initially be taken as shown in Table 3.10.

54

TABLE 3.10 Vehicle Repairs and Road Roughness

Roughness conditions	Spares/repairs factor
Low	1.0
Medium	2.0
High	3.0
Very high	4.0

For vehicle usage of significantly less than 1500 hours per year the spares/repair factor should be multiplied by a utilization factor F

$$\text{where} \quad F = \frac{\text{actual use}}{1500} \text{ (hours per year)}$$

For example, in medium conditions with 1000 hours per year use, the spares/repairs factor would be

$$2.0 \times \frac{1000}{1500} \quad \text{i.e. 1.33 or \$833 per year.}$$

3.7.9 Category descriptions

Gradient—'low' means almost flat roads; 'medium' is roads in rolling terrain; 'high' is in broken terrain; 'very high' means mountainous roads with very steep up and down gradients.
Curvature—'low' means almost straight roads; 'medium' means somewhat winding; 'high' means quite twisting; 'very high' means very twisting roads with almost continuous short bends.
Tyre rolling resistance—'low' means good surface with very little sinkage; 'medium' means a reasonable earth/gravel surface with some sinkage; 'high' means poor surface with quite deep sinkage and ruts; 'very high' means very deep ruts in soft soil, sand, or mud.
Road roughness—'low' means smooth surface; 'medium' is for reasonably graded roads with some ruts; 'high' means considerable ruts or potholes; 'very high' means very rough roads with severe ruts, corrugations or potholes.
Tyre grip—'good' means a good, dry surface; 'medium' is a reasonably good earth/gravel road; 'poor' is for quite slippery roads, probably wet; 'very poor' means very muddy and slippery roads with problems in getting sufficient traction, braking, and cornering grip.

REFERENCES

Crossley, C. P., Kilgour, J., and Morris, J. (1976). SNAIL Transport. Seminar, Lanchester Polytechnic.
Crossley, C. P. (1978). Trends in smallholder mechanization in Kenya. NCAE Occasional Paper No. 8.

Hide, H., Abaynayaka, S. W., Sayer, I., Wyatt, R. J. (1975). The Kenya road transport cost study: research on vehicle operating costs. Transport and Road Research Laboratory Report 672.

Inns, F. M., and Kilgour, J. (1978). Agricultural Tyres, Dunlop Ltd.

International Bank for Reconstruction and Development (1977). An investigative survey of appropriate rural transport for small farms in Kenya.

Morris, J. (1975). A financial/economic feasibility of the SNAIL in Malawi. NCAE.

Robinson, R., Hide, H., Hodges, J. W., Rolt, J., Abaynayaka, S. W. (1975). A road transport investment model for developing countries. Transport and Road Research Laboratory Report 674.

Part 2 *Mechanization Equipment*

4 *Hand and Animal Power*

An ox cart being
used for carrying
domestic water
supply

4.1 HAND POWER

Farming carried out on a hand-tool technology seldom exceeds subsistence levels. The farmer and his family will require to put in their entire effort to producing their food and only in good years will there be any surplus production which can be sold in order to buy goods from outside. An adult human in good health and well fed has a power capability of about 0.07 to 0.1 kW. Specially trained people, such as athletes, can do better than this but only for relatively short periods.

The most commonly used implement is the hand hoe (known variously as Khasu, Jembi, etc.). It can be used in a wide range of soil types with or without a crop or weed cover. The sequence of operation is shown in Figure 4.1, the soil being levered out, as pulling is much more tiring. Adequate back clearance is necessary to reduce frictional losses on the back of the blade as it enters the soil, thus enabling it to work deeper for the same effort. The length of the handle depends on the local tradition but in general long handles and heavy heads give

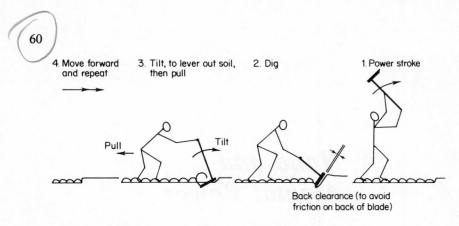

4. Move forward and repeat
3. Tilt, to lever out soil, then pull
2. Dig
1. Power stroke

Pull

Tilt

Back clearance (to avoid friction on back of blade)

FIGURE 4.1 Sequence of operation using hand hoe

deeper work. Spades and forks can also be used but are less suited to hard soil conditions—they are more suitable for irrigated vegetable patches.

The other hand tool widely used is the knife (cutlass, panga, etc.) used for cutting light bush in order to clear the land before cultivating. It is also used to cut the mature crop at harvest time. This tool is not very suitable for clearing larger trees, this being illustrated by the fact that in the traditional shifting cultivation systems the larger trees are left. A local type of axe using a simple triangular-shaped head in a burnt hole at the thick end of the handle is used to clear the larger trees. The type of axe with a hollow metal head into which the shaft is attached by wedges is heavier and more effective but much more difficult for a village blacksmith to make over a charcoal fire.

Giles (1975) established a correlation between available power per hectare and crop yield, Figure 4.2. This indicates a high rate of increase in yield for increasing power inputs up to a level of approximately 0.4 kW/ha (corresponding to a crop

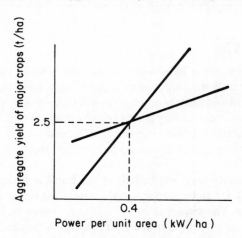

FIGURE 4.2 Relationship between power inputs and crop outputs in agricultural production

output of about 2.5 ton/ha). Typically in Africa one adult works about half a hectare of land providing about 0.1 kW/ha. If 0.4 kW/ha is taken as the desirable level then power supplementation of 0.3 kW/ha is necessary. A pair of oxen would provide each supplementation for about 3–4 hectares of land, about one smallholder farm. Inns (1980) suggests that this is economically the best solution to increasing the productivity of the smallholder farmer but points out that in areas where there is no tradition among arable farmers of animal ownership and care it is extremely difficult to instil into them the necessary management skills and sympathy that the use of animals demands. These are very often the areas where the animal health problems are greatest so that introduction will not necessarily produce the hoped for advantage.

4.2 DRAUGHT ANIMALS

4.2.1 Characteristics of draught animals

Many animals have been used for draught work or transport but the four species most widely used are summarized below together with their advantages and disadvantages (Inns, 1980).

(i) The horse. Advantages:

(1) It is a friendly animal that can become attached to its owner.
(2) It enjoys a certain prestige in Africa, so that the farmer is inclined to give it greater care than he would an ox.
(3) It is easy to train for all kinds of work.
(4) When working, it is fast, easy to handle, docile and can be simply, easily and accurately controlled.

Disadvantages:

(1) It is often too light in Africa, where, under average conditions of maintenance, its weight may be 250–280 kg.
(2) It has a rather weak constitution and needs special care and attention and is costly to maintain.
(3) It is susceptible to trypanosomiasis.
(4) It tires fairly quickly when working.
(5) It is expensive.
(6) Its harness is costly.

(ii) The ox. Advantages:

(1) It works slowly but unflaggingly.
(2) It is hardy and strong and easy to feed.
(3) Its harnessing is simple and the yoke can be made locally at low cost.
(4) Its purchase price is attractive (in some countries one pair of oxen is equal to one horse).

(5) At the end of its working life, it may be sold for meat, and is in fact often sold for this purpose after a period of fattening.

Disadvantages:

(1) It is not a friendly animal, at least so far as many African farmers, who are unaccustomed to the care of animals, are concerned. The farmer considers the ox solely from the point of view of its working value.
(2) The ox needs relatively extensive grazing areas.
(3) It appears to be more difficult to train than the horse and needs more manpower to control it.
(4) When working, it is slow.

(iii) The donkey. Advantages:

(1) It is a friendly, hardy and quiet animal.
(2) It is economical to buy and can be maintained on local farm produce.
(3) It is easy to train and is intelligent.
(4) It is patient when working (light draught work and load carrying).

Disadvantages:

(1) It is very light and limited in strength.
(2) It tires easily if driven too fast.
(3) It is susceptible to trypanosomiasis and harness sores.

(iv) The dromedary. This animal is patient and hardy but difficult to train and sometimes has an awkward temperament.

4.2.2 Choice of breed and individual

It needs to be stressed that in most cases and whenever possible, use should be made of local indigenous breeds. The use of animals transferred within the country or imported animals should be avoided, as they are often not resistant to the local pests and diseases.

(i) Choice of individual. Within a particular species or breed, the qualities to be looked for when choosing a draught animal are as follows:

Conformation. As regards the ox, as a draught animal, it should be powerful, compact, sturdy with well-developed muscles, particularly those of the back and the hindquarters. Its legs should be strong and as short as possible ('low on the ground'). Its chest should be ample and deep.

In the case of draught horses they should have, in addition to the same power characteristics as the ox, short and straight shoulders, long forelegs and muscles as strong as possible (for leverage), sloping hindquarters and short hindlegs.

Character. Whatever the species, a calm and docile animal without vices (tendency to kick or to butt) should be sought.
Age. In the case of the ox, training starts at about two or three years of age; in the horse, at about three years.
Sex. Gelded males are usually used but the mare, the she-ass, and the cow can also be worked.

4.2.3 Utilization and work to be done

The ox is particularly suitable on fairly heavy soil, for relatively deep work (15 cm), for lifting groundnuts, digging-in green manure, pulling loaded carts, and drawing water from wells.

The horse is mainly used on light soil for shallow row-crop work (such as for groundnuts, sorghum, and millet), and for pulling the seeder or the light hoe. No other animal can compare with it for pulling light, shafted carts or cabs but it is not usually strong enough (at any rate in Africa) for ploughing or pulling the groundnut lifter.

The donkey is suitable for light draught work and load carrying, and light work, such as weeding, hoeing, sowing or the use of the single row planter.

The dromedary is used mainly as a beast of burden and, in arid regions, for drawing water.

Table 4.1 shows the work capacity of man and various animals in terms of mechanical power and work output. The data available are somewhat variable, as may be expected for different breeds and for the varying condition of the animals and of the environment in which they are found. The values in the table may, therefore, be used for general guidance and comparison, but should be applied with discretion. The pull and power quoted for animals are based on figures for continuous work output. For short periods they can be greatly exceeded (by a factor of five or more) if compensated by adequate rest periods. The average continuous pull may be taken as about 10–15 per cent of its body weight, but the implement must have at least a five times factor of safety otherwise breakages will be too frequent.

TABLE 4.1 Work Potential of Man and Various Animals

Item	Man	Horse	Ox	Donkey
Weight (kg)	80	400–700	300–900	100–300[a]
Pull (N)	100	500	500	400
Speed (m/s)	1.0	1.0	1.0	1.0
Power (kW)	0.10	0.50	0.5	0.40
Daily work hours (h)	6	6	5	4
Daily work output (MJ)	2	10	9	6

[a] A very wide range of sizes is found in practice.

4.2.4 Energy inputs, power outputs

It is theoretically possible to calculate the amount of food it is necessary to feed the animal in order to get a certain power output; it is the same concept as putting a certain amount of fuel in an internal combustion engine and expecting some power output (except that the 'response time' is far longer with an animal). The following discussion is based on information available for bovines but the principles should be applicable to any animal. Table 4.2 shows typical foods which could be used. From the subsequent calculations and a knowledge of the above crops it would be possible to calculate the amount of land required for growing the crop for the particular animals in question.

TABLE 4.2 Typical foods

Food	Dry matter (%)	Gross energy of dry matter (MJ/kg)
Kikuyu grass	15	5.5
Sudan grass	31	17
Guinea grass	26	7.39
Hay—average quality	85	17
Maize silage	27	18.8
Maize grains	86	19.0
Groundnut cake	90	20.7

This in turn could lead to calculations on the costs and efficiencies of the overall system and enables management decisions to be made, for example is it worth buying in higher quality food to save land so that more cash crop can be grown in order to maximize the farmer's returns? The energy content of any food can be measured in a laboratory using a Bomb calorimeter. This value is called the Gross Energy (GE) and is measured in MJ/kg of Dry Matter. However the animal cannot extract all this energy in its digestive system, partly because of the chemical form of the energy in the food and partly because of the chemical makeup of its digestive system. Digestible Energy (DE) equals 0.45–0.85 GE, 0.45 relating to a poor food (such as barley straw) when the energy is not readily available; and 0.85 relating to a good food (such as barley grains) when most of the energy is available.

Not all the Digestible Energy is available to the animal as there is still some energy in the faeces, urine, and methane produced in the gut, which is rejected in the digestive process. The energy which the animal can make use of is the Metabolizable Energy (ME), which equals 0.8 DE.

The food the animal takes in is used for two purposes, one to maintain the structure, known as the maintenance ration, and the other to provide energy to give a power output. In the case of growing animals a further input of energy is necessary for conversion to a larger structure—muscle, fat, bone, etc. A 1 kg Live Weight Gain (LWG) is equivalent to 28 MJ/day input for beef cows. The energy input for a power output of 1 kW h is equivalent to 3.6 MJ energy. The energy conversion ratio for animals is about 20 per cent if it is a normal healthy adult which means that per 1 kW h output the

ME equals 18 MJ
DE equals 22.5 MJ
GE equals 26.5 to 50 MJ

depending on food type.

The maintenance ration (MR) equals $8.3 + (0.091 W)$ MJ/day, where W is live body weight in kg. The maximum intake per day for the animal depends on the quantity of food it can eat and the energy value of that food. The maximum Dry Matter intake per day is known as the Appetite Limit (AL) and is approximately equal to 0.025 W kg/day.

It is now possible to calculate the amount of food required for a given animal to complete a particular task, and for the farmer to work out the total food requirement per year in terms of tonnes and hence the area of land required and the storage necessary. To illustrate how the quantity of food required can be calculated, the following is given as an example.

Requirement 1 kW hr for 7 hours using bullocks. From Table 4.1 we will require two animals weighing, say, 600 kg each. The Metabolizable Energy in the food required for each animal for maintenance and power would be:

$$8.3 + (0.091 \times 600) + (7 \times 18) = 188.9 \text{ MJ/day.}$$

Taking an average food at 18 MJ/kg Gross Energy with good digestibility, this would provide $18 \times 0.8 \times 0.85$ MJ/kg, that is 12.24 MJ/kg. With poor digestibility it would give $18 \times 0.8 \times 0.45$ MJ/kg, that is 6.48 MJ/kg.
Therefore total food intake required per day is:

with good digestibility $\dfrac{188.9}{12.24} = 15.43$ kg/day

with poor digestibility $\dfrac{188.9}{6.48} = 29.15$ kg/day

Appetite Limit $= 0.025 \text{ W} = 15$ kg/day (Dry Matter).

From the example chosen it can be seen that it is just possible for the animal to eat enough food to give the required output if it is fed on a good digestibility food. On the poor-quality food it would either not produce the work output required or would make up the energy deficit from its internal energy supply (fat). If this output were required for a few days the animal would lose weight and have less stamina and be more susceptible to disease; in any case the work output would fall dramatically. If the animal were fed on good quality hay at 85 per cent Dry Matter then the quantity per day of actual hay would be:

$\dfrac{15.43}{0.85} \times 2 = 36.3$ kg/day for the two animals.

Say the animals work hard for 20 weeks of the year and do nothing useful during the rest of the time, the total food each would require per year would be:

$8.3 + (0.091 \times 600)$ MJ/day, that is 14,152 MJ/year for maintenance only, plus 188.9×140 MJ/year, that 26,446 MJ/year for work; total 40.598 MJ/year.

For two animals being fed on food of 12.24 MJ/kg Gross Energy at 85 per cent Dry Matter, the actual food quantity required is

$$\frac{40598 \times 2}{12.24 \times .85} = 7804 \text{ kg or about 8 tonnes/year for the two animals.}$$

This could be harvested from say $1–1\frac{1}{2}$ hectares of good land.

This area of land could represent one-third to one-half of the land available to the farmer and would consequently be a high 'cost' in his farming system. The situation in practice is not so well defined or controlled and it could be that the animals obtain their maintenance ration from grazing non-production land (i.e. agriculturally unproductive areas such as river edges, roadsides and steep slopes). So the farmer has only to find the 'power' part of the ration.

It should be noted that this section on animal feeding is somewhat simplified but it demonstrates the principles that the food requirements can be worked out scientifically and this should lead to better management decisions. In many cases bought-in concentrates would be used to maintain the correct balance of carbohydrate protein, fat, minerals, and vitamins which is essential for a correct healthy diet.

4.2.5 Harnesses

The attachment of implements to animal is normally achieved by harnessing. The choice of the method of harnessing the animal to the implement is related to the animal's anatomy, simplicity of use, and low cost. Such choice is effectively limited to yokes for bovines and collars or, for lighter work breast harness, for horses and other animals. Collars can be used with bovines, provided they fit high on the shoulders away from the dewlap; a horse collar is not suitable. Head yokes should not be used for bovines, as the forces must be transmitted through the animal's relatively weak neck and therefore its performance is limited.

Figure 4.3 shows some common types of harness. Basically the yoke is a wooden cross-piece to which the animal applies a pushing force from the forehead (head yoke) or neck (neck yoke). The yokes may be either single or double and there are very many regional variations in design. The double neck yoke is the most widely used method for oxen, being simple to make, cheap, and reasonably effective. It will be noticed that the collars and breast bands are usually much better padded than the simple yoke; this spreads out the loads on the animal's skin, thus making it more comfortable and leading to longer work periods, less fatigue, and fewer problems with skin sores.

The accuracy and quality of work performed by draught animals is greatly influenced by ease and effectiveness of control. Ideally the operator should be able

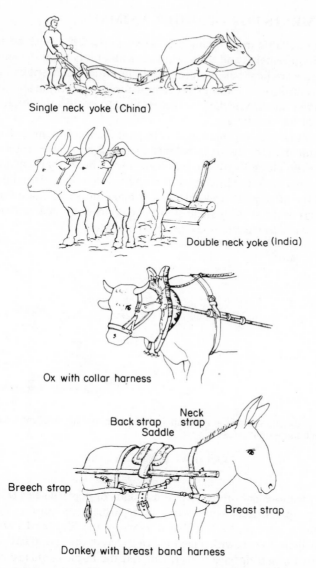

Single neck yoke (China)

Double neck yoke (India)

Ox with collar harness

Back strap / Neck strap / Saddle / Breech strap / Breast strap

Donkey with breast band harness

FIGURE 4.3 Various types of harness (after Inns, 1980)

to guide both the implement and the animal, dispensing with a separate lead man. Control depends on an effective guidance system, good training, and regular practice. For oxen, guidance may be by means of ropes attached to the ears, horns, or nose of the animal. In the latter case the nasal septum must be pierced for the purpose. Uyole Agricultural Centre recommend the insertion of a screwed ring into the nose and have provided details of the training and methods of harnessing found effective at the Centre (Daily News of Tanzania, 1978).

4.3 IMPLEMENTS FOR DRAUGHT ANIMALS

The choice of suitable implements is of great importance both because of the agricultural requirements and the low power availability from the animal. Many modern systems are now based on the minimum tillage concept using the chisel plough and multitined cultivation where moisture conservation and resistance to soil erosion are needed. Mulches are further used to protect the soil from erosion and to control the weeds.

Figure 4.4 shows the various forces acting on the implement and animal. It is possible, as shown in the figure, to balance the soil force on the chisel plough and the weight of the plough with the pull force in the hitch chain by adjusting the length of the chain and the hitch position on the hitch plate. To go deeper move the hitch on the hitch plate upwards and/or make the hitch chain longer. When the plough is running steadily there should be no force at the handles so that the operator has only to steer and control the animals.

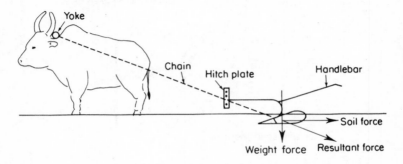

Force in draught chain is in line with (ie the same as) the resultant force of the soil and weight forces

FIGURE 4.4 Implement forces

Many of the more elaborate systems use skids or depth wheels (Figure 4.5) or, better still, proper wheels—tool-frame types, which can convert into carts for carrying quite a large payload (Gibbon, Harvey, and Hubbard 1974). A pair of oxen can easily pull a cart with a payload of 2 tonnes on a good flat surface. With these systems the settings are less critical and the implements are easier to use. However, the load on the skids or wheels should be as low as possible to keep down the parasitic forces—rolling resistance and bearing friction.

4.3.1 Control of implements

Apart from the depth control previously mentioned it is important to be able to steer the implements accurately, especially to facilitate subsequent interrow weeding. Single-point implements can easily be steered by tilting the implement slightly. This displaces the line of pull sideways and makes the implement drift

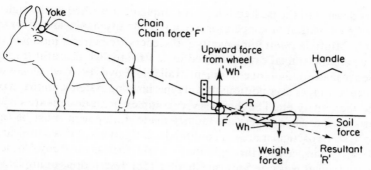

Yoke

Chain
Chain force 'F'

Upward force
from wheel
'Wh'

Handle

R

Soil
force

F   Wh

Weight
force

Resultant
'R'

Force in draught chain and resultant force not in line, the 'difference' is
taken up by force on depth wheel

FIGURE 4.5 Implement forces using a depth wheel

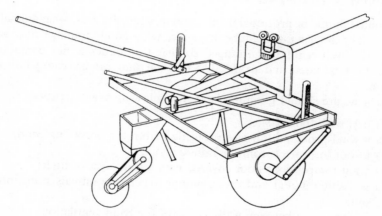

Toolbar with wheels (Makonatsotlhe – Botswana)

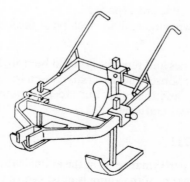

Toolbar with skids (Ariana)

FIGURE 4.6 Common types of toolbars (after Inns, 1980)

slightly sideways to the desired line. If the implement is tilted to the right it will steer to the left without the operator having to put any significant effort into the operation. Multiple-point implements and seeders are better if mounted on tool frames where the animal can be steered to control the whole outfit (Figure 4.6).

There are many designs of equipment available (Boyd, 1976) which are suitable for particular ranges of conditions of soil type, farming systems, animal type, level of economic activity, etc. The choice is very complex; some designs are suitable for local production while others are factory-made. Equipment must be carefully selected to suit a particular area or country. Investigations of this nature are being carried out at many centres throughout the world, such as the Kenya Agricultural Machinery Unit at Nakuru. See also Chapter 7 for descriptions of implements, as many of the smaller, simpler, and lighter designs can be used successfully with animal power.

4.4 ANIMAL HOUSING

The animal should be protected from sun, rain, wind, diseases, other animals, and thieves. The housing should be hygienic, with good drainage and general cleanliness, well-ventilated, arranged so that the fodder is not contaminated before it can be eaten, and arranged so that the dung can be readily utilized for manure.

The structure should generally have the following characteristics:

  (1) dry, well-drained site;
  (2) a wind-proof wall to give protection from the prevailing wind;
  (3) a roof to protect from the rain and sun;
  (4) a hard impervious floor covered with a thin layer of litter;
  (5) walls high enough and strong enough to keep the animals in and intruders out;
  (6) a large space between walls and roof for good ventilation;
  (7) a water trough;
  (8) a food manger;
  (9) a food store of sufficient capacity;
(10) a dung heap down wind and down slope of the animal house and human habitation.

The simple farm medicines and remedies should be stored in a safe, secure place, probably in the farmer's house. The materials chosen for the structure depend on what is available locally and the price, but excellent housing can be made from local materials using local craftsmen.

4.5 ANIMAL HEALTH

In order to get the best performance from the animals they must be kept in good health. The person in charge of the animals must keep an ever watchful eye open for any signs of abnormality which would give an indication of some disease. If in doubt, call the veterinary officer.

When the animal is ill it will show various symptoms or signs; some of the more common symptoms are as follows:

(1) Appearance—a diseased animal will often have an abnormal posture and be restless.
(2) Behaviour—the animal will be generally unresponsive and lifeless; some disorders, however, may make the animal aggressive.
(3) Appetite and feeding—the first sign of some diseases is often that the animal will not feed normally: it may be reluctant to feed, have difficulty in swallowing or chewing. Some diseases, such as gut parasites, may cause it to feed excessively. Unnatural ingestion of non-food substances, such as eating soil, licking rocks and metal, may indicate various deficiencies in the diet.
(4) Defaecation—excessively watery or hard faeces often indicate that there is some disease present. Blood-stained faeces are a symptom of some diseases.
(5) Urination—abnormal colour or consistency are often signs of disorder.
(6) Skin and coat—a healthy animal has a warm, dry, soft, pliable skin and coat; a scaly cracking skin with sores and loss of coat is usually a bad sign.
(7) Mucous membranes—these are at the mouth, nose, eyes, anus, and reproductive openings and should be generally moist and pink with no excessive discharges or pus exudates.
(8) Temperature, pulse rate, and respiration rate—healthy animals show these characteristics within certain ranges (Table 4.3), and radical changes are usually a sign of some disease.

TABLE 4.3

Animal	Pulse rate per minute under jaw bone	Respiration rate per minute	Temperature °C
Cattle	40–50	10–30	38.5–39.5
Horses	36–42	8–16	37.5–38.5
Donkies	40–60	40–56	36.0–38.0

TABLE 4.4 Common Diseases of Draught Animals. (Reproduced by permission of Macmillan Books Ltd)

I *Infectious diseases*

Disease	General	Symptoms	Control
Rinderpest	1. A highly contagious and infectious disease of cattle. It is caused by a virus which mainly attacks the mucous membranes of the alimentary tract (from mouth to rectum) 2. In East Africa cases are becoming rare with routine vaccinations 3. This is a killer disease and no chances should be taken with it	1. Higher fever over 40°C; severe dullness and loss of appetite 2. Profuse diarrhoea and blood-stained faeces 3. The mouth, nose, and muzzle will be hot with fast breathing 4. Lachrimation and nasal discharge 5. Rapid dehydration resulting in emaciation with sunken eyes	1. No treatment 2. Affected animal plus the whole herd should be slaughtered 3. Application of quarantine whenever there is an outbreak 4. Vaccination of all cattle from about one year-old every year. The immunity lasts 3–6 years.
Foot and mouth	1. A highly contagious and infectious disease of cattle, sheep, goats, and pigs 2. Caused by a virus which attacks the mucous membranes of the mouth and coronet 3. The disease is very severe in cattle and pigs but mild in sheep and goats 4. Foot and mouth is an endemic disease in East Africa. It is a killer for exotic cattle but the Zebu are fairly resistant and many do recover 5. If strict measures are taken as with rinderpest it can be eradicated	1. Fever, dullness and loss of appetite 2. Profuse and continuous salivation in the mouth 3. Lameness due to wounds in the coronet on all legs 4. Loss in milk production and emaciation. Death may result or animal gets thin for several weeks until recovery 5. Wounds or blisters on tongue, gums, and palate	1. No treatment 2. Affected animals should be slaughtered 3. Application of quarantine whenever there is an outbreak 4. Regular vaccinations every six months

Rabies	1. An infectious disease of mammals caused by a virus which attacks the nerve cells. Wild canines and dogs act as carriers and transmit it with their bites 2. The disease is rare even in dogs except in areas bordering game parks	1. In domestic animals and man the major sign is madness, i.e. chasing other animals, biting objects, and running aimlessly 2. Death results after some weeks	1. Annual vaccination of dogs 2. Killing of any dogs suspected 3. Killing any animal suspected 4. No treatment
Pneumonia	1. An infection of the lungs (lung tissue and pleura) 2. Can be caused by several types of bacteria or viruses included in worms and dust particles 3. Affects all mammals and birds 4. Pneumonia is a particularly serious disease in calves and young animals, especially under dirty, damp, or dusty conditions. Keep buildings well-ventilated, clean and warm	1. Difficult breathing and coughing 2. Temperature may be high or low 3. Nasal discharge and congestion of bronchioles 4. Animal reluctant to move, dull and sleepy 5. Loss of appetite and dullness	1. Treat early cases with antibiotics 2. Nurse in a warm shelter 3. Provide soft feeds and water
Anthrax	1. An acute infectious disease of mammals (cattle, sheep, goats, pigs, and humans) 2. Caused by the bacterium *Bacillus anthracis* 3. Carnivores and birds are less susceptible	1. High fever of over 40°C. 2. Shivering, loss of appetite and dullness 3. Dysentery (blood-stained faeces) 4. Sudden death in cattle within 24 hours 5. In pigs and sometimes humans a mild form with big swellings occurs 6. A dead animal will show the following: (a) Dark watery blood flowing from anus, vulva, mouth and nose (b) An excessively blown-up stomach (c) Absence of rigor mortis	1. Treat early with antibiotics 2. If dead dispose of carcass by complete burning or bury 3 m deep 3. Vaccinate animals every year 4. Never open a carcass that shows symptoms of anthrax 5. Never eat meat from an animal that has died suddenly. 6. Report suspected anthrax cases to the veterinary authorities as soon as possible

Disease	General	Symptoms	Control
Colibacillosis or Scours	1. An infectious disease particularly of young animals, calves, piglets, sheep, and goats and also human beings, before weaning (milk-feeding stage) 2. The major sign in this disease is excessive diarrhoea termed scouring. The faeces are usually pungent smelling	1. Profuse and smelling diarrhoea 2. Dullness and loss of appetite 3. Slight rise in temperature 4. Sudden deaths in piglets, with stomach blown-up and hard	1. Treat with antibiotics 2. Strict cleanliness must be observed in calf pens, pig houses and sheep/goat pens. Avoid damp, wet conditions
Blackquarter	1. An acute infectious disease of ruminants (cattle, sheep and goats) 2. Caused by *Clostridium chauvei* and *Clostridium septicum*	1. High fever 2. Shivering, no appetite, and dullness 3. Lameness—usually a pronounced limp. Muscle is swollen and very painful	1. Treat early cases with antibiotics 2. If dead dispose of carcass as in anthrax 3. Vaccinate animals yearly 4. Never open a carcass that shows symptoms of blackquarter
Mastitis	1. An infectious disease that can be acute or chronic, affecting the mammary glands of mammals (cattle, buffalo, sheep, goats, pig, bitch, woman) 2. Caused by many types of bacteria, but mainly *Streptococcus* and *Staphylococcus* groups	1. Milk contains pus or blood, turns watery or comes in thick clots 2. Udder and teat swollen, animal rejects suckling or milking and kicks 3. Death may result or the affected udder quarter dies and gives no milk	1. Treat early cases with antibiotics 2. Apply infusion of long-acting antibiotics at drying 3. Milk out teat and massage with hot water and give antibiotic through teat canal 4. Strict cleanliness and use of disinfectants during milking 5. Use the right milking technique

Disease	Description	Signs	Treatment/Control
Foot rot	1. An infectious disease attacking the hooves of all animals 2. Caused by bacteria of the *Fusiformis* group 3. In foot rot there should be more than one foot affected: if it is only one foot, then it is most likely an abscess 4. The disease is particularly common during wet weather and in wet areas. It is rather rare in dry areas or during drought	1. Swollen painful hoof, making animal go lame 2. Parts of hoof may contain pus and smell rotten	1. Treat with antibiotics 2. Trim properly and remove the affected part if rotten, then isolate the animal 3. Provide cattle, sheep, and goats with a foot bath of copper sulphate solution every week 4. Routine trimming and examination of feet
Brucellosis	1. A contagious infectious disease affecting cattle, sheep, goats, pigs, and man. 2. Caused by *Brucella* bacteria 3. This disease does not kill animals except in the foetal stage, but it is very serious for human beings	1. In animals mainly abortion followed by a brownish vaginal discharge 2. Retention of placenta 3. Swollen testicles in rams 4. In the case of sows many piglets are born dead 5. Apart from these signs the animal does not appear sick	1. Treatments are not satisfactory since carriers may result 2. Cull and slaughter the affected animals 3. Vaccinate all young females below 12 months, especially cattle 4. Do not touch aborted foetuses or remove placentas without gloves 5. Milk from infected animals must be boiled before drinking
Coccidiosis	1. A protozoan disease of poultry, calves, rabbits, kids, and lambs. 2. Caused by organisms of the *Eimeria* spp. 3. The organism attacks the linings of the alimentary tract (small and large intestines and liver)	1. Diarrhoea, dysentery, and emaciation in all animal species 2. Rough feathers, dullness, and drooping wings in poultry 3. Sudden deaths in rabbits and kids 4. Aparts from these, the symptoms are vague and general	1. There are many types of drugs (coccidiostats) available for treatment and prevention 2. For prevention these drugs are often given in drinking water or feed for poultry and rabbits

TABLE 4.5 Vector transmitted diseases

Disease	General	Symptoms	Control
Trypanosomiasis 'Nagana'	1. An infectious protozoan disease of domestic animals (cattle, sheep, goats, dogs, pigs, horses, and man). It is transmitted by the tsetse fly 2. This disease is easily eliminated with eradication of the tsetse fly. However a large area of Africa between the tropics is infested by tsetse flies, except the highlands, and therefore this disease will not be eradicated in the near future	1. Fever, dullness and loss of appetite 2. Marked anaemia resulting in licking soil and emaciation 3. Swollen lymph nodes 4. Running eyes which leads to blindness 5. Death may occur after several weeks	1. Many trypanocidal drugs are available, e.g. Berenil, Antricide, Samonin, etc., used for prevention and cure. 2. Control of the tsetse fly by bush-cleaning and application of insecticides
II *Tick-borne diseases*			
Heart water is a rickettsial disease of ruminants (cattle, sheep, and goats)	Ticks can be effectively controlled by the following: (a) Fencing and rotational grazing (b) Burning of old pastures and grass (c) Ploughing and harrowing in dry season (d) Regular use of acaricides in dipping or spraying	1. Fever, of 39–41°C, dullness and loss of appetite 2. Protrusion of the tongue 3. Animal moves in circles and becomes restless, placing head against hard objects, twitching eyelids 4. When it falls, the legs keep paddling in the air	1. Treat early cases with tetracyc-line antibiotics and sulpha-dilimidine 2. Tick control is still the best method

| ECF is a protozoan disease affecting only cattle. Also called Coast fever. | Transmitted by the brown eartick (Ripicephaelus appendiculatus) | 1. High temperature, salivation, and lachrimation. Animal gets thin very quickly and dies in 2–3 days
2. Petechiae or small haemorrhages in vulva and oral mucous membranes | 1. No medicine has been discovered to cure it
2. Control of ticks by spraying and dipping |
| Anaplasmosis and Redwater are also protozoan diseases transmitted by ticks. They affect only cattle | | 1. Fever, constipation, and dullness
2. Anaemia-yellowish mucous membranes. Animal licks soil
3. Swollen lymph glands
4. Red urine (like strong coffee) is seen in the case of Redwater | 1. Some response is reported with tetracycline antibiotics and Berenil, however a number of cases die.
2. Tick control is still the best method |

D. N. Ngugi, P. K. Karau, and W. Ngugo (1978). *East African Agriculture*. Reproduced by permission of Macmillan Books Ltd.

REFERENCES

Boyd, J. (1976). *Tools for Agriculture* (Intermediate Technology Publications).

Daily News of Tanzania (1978). Towards draught animal mechanization, *Daily News*, 1 June, p. 4.

Gibbon, D., Harvey, J., and Hubbard, K. (1974). A minimum tillage system for Botswana, *World Crops*, September–October, 229–34.

Giles, G. W. (1975). The reorientation of agricultural mechanization for the developing countries. FAO/OECD report of expert panel (FAO, Rome).

Inns, F. M. (1980). Animal Power in Agricultural Production Systems, *World Animal Review*, **34**, 2–10.

Ngugi, D. N., Karau, P. K., and Ngugo, W. (1978). *East African Agriculture* (Macmillan).

5 Characteristics of Small Engines and Transmissions

A small diesel engine and simple, belt drive transmission provide the basis of a low cost transporter

5.1 ENGINES

An engine is a device which, when provided with fuel and air, produces rotary power. The efficiency of the conversion of chemical energy in the fuel to mechanical energy at the engine output is not high (maximum efficiencies are usually in the range 25–35 per cent for petrol engines, and 30–40 per cent for diesels). The principle of most internal combustion engines is to convert a reciprocating thrust, exerted by burning fuel on a piston, into a rotary motion at the engine output. In order to achieve the power output it is necessary to have quite a complex series of mechanisms and ancilliary systems (such as lubrication, cooling, fuel supply, filtration, exhaust, electrics, etc.), and in order to get a reasonable degree of smoothness, particularly with big engines, it is usual to have a number of cylinders (2,3,4,5,6,8), which increases the complications.

The modern internal combustion reciprocating engine, therefore, is in engineering terms a complex, noisy, costly, and inefficient device for producing power. Having said that, however, there are two major advantages associated with it. The first is that it is now almost universal, to be found operating in virtually every part of the globe. The second is that it is very adaptable. Provided that it is in satisfactory condition, an engine requires only a 'can of fuel' to make it operate, and it can in principle be used for stationary, marine, aeronautical, automotive, or agricultural operations.

Up to recently the provision of the 'can of fuel' mentioned above has been the least of the problems. Maintenance facilities, operator skills, repairs, and spare parts have been (and still are) a major headache associated with operating internal combustion engines at the smallholder level. In recent years, of course, the can of fuel has come to be a major problem in itself. A number of developing countries now spend an alarming proportion of their foreign exchange on the purchase of imported oil-based fuel supplies. The known limits of these supplies are likely to produce rising (real) fuel costs from these sources until they are exhausted. Consequently a point of view held in some quarters is that the internal combustion engine is doomed. The view taken in this book is that it is not. This view may need expanding. A major reason why engine-powered technology (cars, trucks, buses, trains, boats, tractors, bulldozers, graders, etc.) has spread to such a large extent worldwide is that it provides a different order of magnitude of power output compared with a human worker. An average man in good health is able, throughout a working day, to provide an output of 0.07 kW (one-tenth of a horsepower). Even a very small engine can produce twenty or thirty times this power, and conventional truck/tractor engines (say 70 kW) produce no less than a thousand times this power output.

There is, in short, nothing to match the internal combustion engine as a convenient and universally available source of power. It is felt, therefore, that more effort is likely to be directed towards finding alternative fuels for existing engines than towards finding alternatives to the engine itself. Certainly, solar, wind, tidal, nuclear, geothermal, and biomass based energy sources will be likely to increase but for remote or mobile applications (including transport and field work in agriculture) it seems likely that more efficient internal combustion engines, probably running on fuel derived from coal, sugarcane, cassava, sunflower oil and similar sources, will remain a major supplier of power into the next century. This point is discussed further in Section 5.1.5.

5.1.1 Performance characteristics of engines

Engines produce power based on a simple formula:

$$\text{Power (kW)} = \text{torque (kNm)} \times \text{speed (rad/s)}$$

Thus if an engine were able to provide a constant torque, the power produced would increase linearly with speed (Figure 5.1). In practice the torque curve of a typical engine is not flat, and neither does it extend back to zero speed (Figure 5.2). There is a limit in speed below which an engine is not designed to produce useful work. This speed varies with engine size, but is usually in the region of one-third maximum speed (i.e. 800 revs/min for an engine rated at 2600 revs/min, and 1300 revs/min for a 4000 revs/min engine).

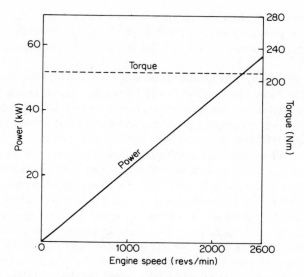

FIGURE 5.1 Performance characteristics of a constant torque engine

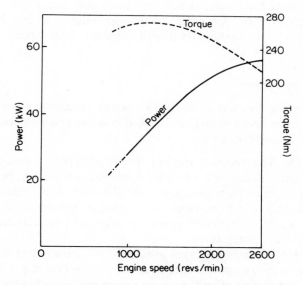

FIGURE 5.2 Performance characteristics of a practical engine

The non-flat torque curve is advantageous since it provides a 'back up' of torque in the event of an increased load occurring on an engine running at high speed. (The speed, and hence the power, will drop in the process but at least the engine will not 'die' immediately.) As power is based partly on torque, the actual power curve will, therefore, be of the form shown in Figure 5.2. It should be noted that the maximum power can only be produced at or near maximum speed. (The performance curves shown are for a medium-sized agricultural diesel engine.)

The values of power, torque, and speed will vary considerably over a range of engines, but the basic characteristics will be similar. Upon examination of the output characteristics of various engines, in relation to the requirements of the machines they power, a common problem can be identified. Most engines will not run well below 1000 revs/min and will not develop full power below 2000–3000 revs/min. When used on a self-propelled machine such as a tractor or transport vehicle, the major function of the engine is to provide power to the wheels. However, a tractor rear wheel of 1.5 m diameter, if driven at full engine speed of say 2200 revs/min, would cause the tractor to travel across the field at a speed of 622 km/h. This is not a convenient speed for field operations.

It may be concluded, therefore, that where an engine is to power wheels for forward motion there is nearly always a considerable mismatch between the output speed characteristics of the engine and the input speed requirements of the wheel. This mismatch is one of the major reasons for the complexity of a tractor transmission system (dealt with later). It may be noted here that the only type of engine which need not possess this speed mismatch is the steam engine. This may be able to produce full torque at zero speed and may run at only 200 revs/min. The high initial cost and low efficiency of steam engines means, however, that they are probably not worth considering in most circumstances.

5.1.2 Types of engines

Although engines do exist in almost all power categories from about 1 kW upwards they may conveniently be classified into power classes, each tending to have distinct characteristics.

(1) 1–10 kW. These engines may be either diesel or petrol (usually the latter in the smaller sizes), will usually be single-cylinder, air-cooled and will operate at maximum speeds of 3000–4000 revs/min. Usual applications are for stationary equipment such as pumps, generators, and processing plant, and for small self-propelled machines such as mowers, small tractors, and dump trucks.

(2) 10–20 kW. These engines are usually diesels, have two or three cylinders, are still often air-cooled, operate at about 3000 revs/min and are used for mills, pumps, and other stationary equipment, as well as for some slightly larger unconventional small tractors, often based on 'dump truck' technology.

(3) 20–50 kW. There are two distinct types within this category. The first is the automotive petrol engine as fitted to motor cars. These operate at around 5000 revs/min but are not usually available for other applications and may therefore be discounted here. The second type is the medium

diesel engine, having three or four cylinders, water-cooling, and operating at about 2500 revs/min. These are typically fitted to medium-sized tractors (three cylinders) and trucks (four cylinders). They are also available in the form of 'industrial' engines for various applications such as stationary power.

(4) 50–100 kW. Probably outside the sphere of smallholder mechanization, these engines are included for completeness. Neglecting the automotive petrol engine again, most of the characteristics of these diesel engines are similar to those described in (3), the extra power being provided by increasing the number of cylinders to six or eight and/or turbo-charging the engine. Applications are large tractors, trucks, and earthmovers.

5.1.3 Temperature and altitude effects

Engines are rated at near sea level when running in cool air. These conditions vary relatively little in the countries in which most of the engines were originally designed. When operated in many developing countries, however, altitude and air temperature effects can give rise to quite a significant drop in engine power. The reduction normally quoted (BS 649) is

Altitude—$3\frac{1}{2}$ per cent per 300 m above 150 m
Temperature—2 per cent per $5\frac{1}{2}°C$ above 30°C

It can therefore be seen that, if a tractor is working at 1500 m in an air temperature of 40°C, the reduction in power is 15.75 per cent plus 3.64 per cent = 19.4 per cent. A 50 kW engine would therefore be reduced in power to 40 kW unless provided with a turbo-charger (which could eliminate the losses). The reason for the power loss phenomenon, incidentally, is that warm air is less dense, as is air at lower pressure (i.e. at altitude). The available oxygen, on which combustion depends, is therefore reduced.

5.1.4 Fuel consumption

The fuel consumed by an engine depends on three main factors:

(1) The consumption characteristics of the engine per unit of power used, expressed as 'specific fuel consumption' in litres per kilowatt hour (l/kWh).
(2) The amount of power being used (in kW, from zero to maximum).
(3) The condition of the engine compared with its original specification.

It is therefore not meaningful to attempt to calculate the fuel consumption of a given engine in terms of litres per hour unless the proportion of the power being used is known, that is fuel consumption (l/h) = specific fuel consumption (l/kWh) × power used (kW).

At full power the specific fuel consumption (SFC) of various engines can be taken as shown in Table 5.1. Therefore a 5 kW engine with an SFC of 0.3 l/kWh would use 5 × 0.3 = 1.5 l/h at full load. But would a 10 kW engine with the same

TABLE 5.1

Type of engine	Rate of fuel use at full power (l/kWh)
Automotive diesel	0.29
Medium diesel (5–20 kW)	0.32
Small diesel (> 5 kW)	0.35
Automotive petrol	0.38
Small petrol (> 10 kW)	0.50

SFC, loaded only to 5 kW, consume fuel at the same rate? The answer is—it depends. The part load performance of engines is usually not as good as at full load, but the way in which the load is applied now becomes important. Running at full speed and half torque (to give half power) an engine will consume more fuel—up to 130 per cent of the normal SFC—than when running at full torque and half speed (also giving half power). This is because of the characteristics of the performance 'map' of an engine. An example, taken from Matthews (1982), is given in Figure 5.3. Since it is not usually known whether a machine operating at part load is at full speed or at full torque, it is reasonable to take a mean between the two conditions. A convenient and simple formula is to modify the SFC at part load by a factor F given by

$$F = \left(\frac{\text{power used}}{\text{full power}} \right)^{-0.15}$$

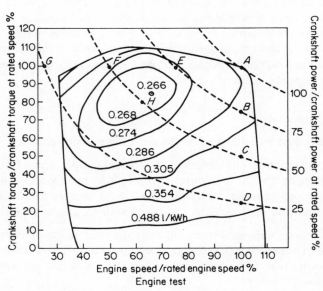

FIGURE 5.3 Fuel economy measured over operating range of an engine (after Matthews, 1982, reproduced by permission of the Institution of Agricultural Engineers)

Operating at half power then gives an F value of 1.109 so that the 10 kW engine in the example above would have an 'effective' SFC of $0.3 \times 1.109 = 0.33$ l/kWh and the fuel consumed at half load would then be $5 \times 0.33 = 1.66$ l/h.

This is a reasonable approximation in practice. In detail, using as an example the engine map shown in Figure 5.3 it can be seen that at point A (full power, full speed) the specific fuel consumption is about 0.29 l/kWh. At full speed but three-quarter power (point B) the SFC is 0.296 while at full speed half power (point C) it is 0.33. When the power consumed has fallen to a quarter full power (but still at full speed) the SFC is 0.427 l/kWh (point D), that is it is up by 47 per cent on that at full power.

What is happening here is that *speed* is being kept constant (at full rated speed) while the amount of *torque* being used is reduced to give reduced power. It is more economical to keep torque as high as possible and decrease speed. In this way the SFC at three-quarter power (point E) is down to 0.273 and at half power (point F) it is 0.28 l/kWh. (It is not possible to get the speed down to a quarter full speed at full torque, in order to get a quarter power, because the resultant point G is 'off the map'.)

In terms of utilizing part power efficiently it therefore makes a difference whether the engine is run at high speed and lower torque, or the converse. For example, with a 50 kW engine at half power (points C and F) the amount of fuel used in these two ways would be 8.25 litres per hour and 7 litres per hour respectively. Over a 2000 hour operating year the amount of fuel saved by operating at point F instead of C would be 2500 litres. It should be noted that the engine would most efficiently be operated at point H, which also gives half power (at 80 per cent torque and 62.5 per cent full speed). Here the SFC is down to below 0.267, the fuel per hour would be 6.67 litres and the saving in a year could be 3150 litres of fuel.

5.1.5 Alternative fuels

The view was expressed, in Section 5.1 of this chapter, that the internal combustion engine, both spark ignition ('petrol') and compression ignition ('diesel'), is likely to survive in substantially its present form for a considerable time to come. Alternative forms (steam, turbine, rotary, etc.) have not made a great deal of impact. In the search for better fuel consumption of vehicles, certain trends are apparent. Reduced weight, smaller-sized engines, smaller frontal area, and lower aerodynamic drag factors, together with higher gearing have made substantial improvements to road vehicle fuel performance. However, some other factors, such as emission control, have tended to increase weight, complexity, cost, and fuel consumption of motor car engines.

Most agricultural machines and vehicles are too slow-moving to be concerned with aerodynamics and it may be thought that the relatively low engine 'population' (except in capital cities) in developing countries, together with the need to reduce the cost and complexity of equipment, will mean that environmental factors such as emissions may be regarded as less important. There is however no reasonable hope that existing engine designs can be improved sufficiently to make a substantial difference to the amount of petroleum-based fuel requiring importation into most (non-producing) developing countries, or even to keep pace with the rate of fuel price increases.

Increasing efforts are therefore being made to develop alternative fuels which can be used either to supplement petroleum fuels (i.e. by mixing) or to replace them altogether. Engines can now be designed to run entirely on fuels such as hydrogen or alcohol. Standard petrol engines will run reasonably well on mixtures of alcohol and petroleum fuels. Sunflower oil is regarded as an important possibility to substitute for diesel fuel, along with other vegetable oils. The total potential of sources for non-petroleum-derived fuels is therefore considerable and includes corn, sugar cane, sugar beet, coal, solid wastes, industrial byproducts, agricultural wastes, forest wastes, natural gas, cassava, palm and palm kernel oil, soya beans, sorghum, rapeseed oil, and molasses.

There are two areas of activity which, in any case of possible substitution of fuel, will need consideration. The first is the agronomic/economic question. Any biological fuel source will require land cultivation, planting, tending and harvesting operations, transport, and processing. The total fuel output from such a set up must clearly be greater than the fuel equivalent cost of the inputs. In some cases, such as sugar cane producing alcohol in Brazil, the balance appears satisfactory. It is also stated that cultivating sunflower crops in some conditions requires an input of only one-tenth of the fuel that can be achieved from processing the crop. Each situation will, however, vary and it would be a mistake to imagine that a suitable alternative fuel can easily be found for all conditions.

The second factor requiring consideration is a technical one. Internal combustion engines have been developed over three-quarters of a century to run either on diesel fuel or on petrol. The characteristics of most alternative fuels possess many similarities to diesel or petrol but also some differences. Where engines are run completely on new fuels, problems may occur with cold starting, high fuel consumption, and reduced service/maintenance life. Running diesel engines on vegetable oils, for example, tends to result in considerable coking of injectors (Bacon, Bacon, Moncrieff, and Walker, 1982). Similarly, standard petrol engines run on a gasoline/alcohol mixture may display coking-up of the spark plugs. Finally, it should be mentioned that research findings obtained under carefully controlled conditions may not be duplicated when extended into widespread usage, when the quality of the fuel supply (due to problems such as water absorption by alcohol fuels) can be lower than desirable. Nonetheless, human nature has a remarkable history of adapting to enforced change. With the wide variety of possible alternative fuels being considered, it is likely that chemical, economic, and mechanical changes will be made; under the threat of dwindling fossil fuels the incentives will certainly become increasingly strong.

5.1.6 Fuel and air filters

An engine maintained in good condition should give performance satisfactorily close to specification, provided that it is supplied with two basic commodities—

clean fuel and clean air. Both of these can be difficult to obtain in smallholder conditions and a good deal of the 'remedial' action is normally taken by the fuel and air filters fitted to the engines. However, these will only be effective if maintained in good condition and prevented from blocking. The time between maintenance checks varies with filter design and condition but, for example, an engine with a small-capacity, oil-bath air cleaner operating in dry season dusty conditions may need filter cleaning *daily*. This is an extreme case (but one experienced by the authors). Usually a weekly clean and replacement of the oil in the cleaner is sufficient.

It is important that this maintenance is not overlooked. Oil and grit form a compound similar to valve grinding paste and, if allowed into the engine in any significant quantity, will produce a similar effect, on expensive components rotating or reciprocating at up to 50 times per second. Dust penetrating into the crankcase through dip stick seals or crankcase filters can also give rise to wear in bearings and oil control rings, due to dust absorption by the lubricating oil (Stevens, 1982).

Fuel filters also need cleaning frequently, although in the case of fuel it is easier, by good management, to ensure that the fuel is maintained in as clean a state as possible while awaiting transfer to the fuel tank of the machine. Refuelling is best performed after work to reduce water vapour problems. Siphon tubes, filler hoses, and tank caps must be kept free of dirt. Engine filters can be relied on to clean the fuel only until the contamination (by water or dirt) is sufficient to block the filters and stop or damage the engine.

5.1.7 Design life

In engineering terms the expression 'design life' can be taken as the time in hours during which major components are estimated, at the design stage, to be likely to continue to function fairly close to their specified performance with normal maintenance. In the case of a rolling bearing, for example, it is usually selected with a certain B_{10} life, meaning that 90 per cent of a large sample will still be functioning at the end of this period.

Two associated problems arise with machine selection and operation. The first is that it is not usually possible to determine what the design life decision originally was. Handbooks from some manufacturers of small engines, for example, introduce their products as giving 'many hours of reliable service'. Others ignore the question altogether. From the maintenance instructions of others it is possible to detect the kind of hours running the designers had in mind between major overhauls. Optional extras such as spares packs for 2000 and 4000 hours operation are encouraging signs with a small engine.

The second problem is that the actual operating conditions may not correspond to those considered as reasonable by the designer. Major potential problem areas are mentioned in Section 1.2.3. Other examples are: low operator skill; blockage of air-cooled engines with vegetation trash; lack of spare parts; inadequate maintenance; tampering; and unskilled repair.

A further factor associated with operation at the smallholder level is that small machines or components may be selected on price and nominal performance, rather than on longevity. Many small petrol engines, for example, are not really designed to produce continuous full power throughout their working life. When used on a mower or small garden rotavator the combination of low power demand and low total annual use can mean that the engines will last for several years. When used at full power in smallholder conditions, however, the expected life of these engines may not extend beyond about 1000 hours, with some considerably less. A normal working year (40 hours per week) approaches 2000 hours, and absolute 'full time' operation 24 hours a day (as might be required when driving pumps, etc.) will reach 1000 hours in less than six weeks.

5.2 TRANSMISSION SYSTEMS

Consider the case of a tractor, which is designed to produce a drawbar pull F at a forward speed V. We know that the drawbar power required is then P

$$\text{where} \quad P(\text{kW}) = F(\text{kN}) \times V(\text{m/s})$$

Given a certain value of available power, we can in theory choose whether we have a large pull and a low speed, or vice versa.

This can be shown on a pull/speed curve (Figure 5.4 with the available power drawn in the form of an 'envelope'. Provided we operate in the area beneath the power envelope we can choose how we make use of the available power. For example, operating at point A gives us 1 kN of pull at 10 m/S and fully consumes the available 10 kW. Conversely, operating at 1 m/s could provide a pull of 10 kN using the 10 kW available power (point B). Are these operating points possible and desirable in practice? Certainly they are both desirable. Point A (10 m/s or 36 km/h) is a useful transport speed although, because of the low pull available, the hill-climbing ability of the machine would not be great (with a

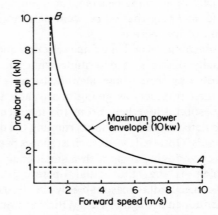

FIGURE 5.4 Maximum available power 'envelope' for an engine

1 tonne machine on a good tar road, about 5° of slope). Point *B* is also a useful possibility, as two mouldboard, disc, or chisel ploughs could be pulled even in hard soil with the 10 kN pull available. Most tractors however, will pull only about 65 per cent of their weight in average conditions. To pull 10 kN at a reasonable slip would therefore require a 1.6 tonne tractor at least (with an engine power of 13.5 kW to allow for losses). This would represent a rather larger than usual 'small tractor', but is certainly a possibility. Even where the full pull cannot be developed it may be useful to have a low forward speed available for easy tractor control in crops. The conclusion here, then, is that points 1 and 2 in Figure 5.4 probably represent either end of a reasonable scale of operation, each using full available power but producing very different characteristics.

In practice, it is almost always necessary to have intermediate points of operation available as well, for example 5 kN at 2 m/s, 2 kN at 5 m/s and so on. These points are achieved by the use of intermediate ratios in the transmission system. Figure 5.5 shows the theoretical characteristics of an actual small/medium tractor with six gears. For a transport vehicle the range would be moved

Tractor

Max. engine power 33 kW 2250 revs/min
Weight 1450 kgf (14.2 kN)
Tyres 10–28

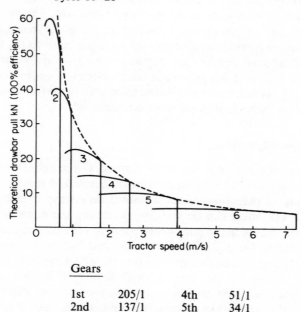

Gears

1st	205/1	4th	51/1
2nd	137/1	5th	34/1
3rd	75/1	6th	19/1

FIGURE 5.5 Practical available power output from a tractor

upwards along the speed axis, to give much higher speeds in top gear (say 25 m/s or 90 km/h) and also in low gear (perhaps 3.6 m/s or 13 km/h).

It can thus be seen that given a certain engine power it is the design of the transmission system which dictates whether the machine will be useful for traction or transport. The important factor is the difference in rotational speed between the engine and the wheels, achieved by a certain overall reduction in the transmission system.

Let us now return to the problem identified in Section 5.1.1 where a tractor was found to travel at an excessive speed in the absence of such a reduction. The tractor in question would probably need a top gear speed for transport of about 30 km/h. At full engine speed this requires a reduction (between engine and wheel) of 20.7 to 1. In bottom gear the maximum speed would need to be about 3.3 km/h, requiring a reduction of 188.5 to 1.

This brings us to the conclusion that a transmission system needs to perform three functions:

(1) To *connect* the engine to the drive wheels.
(2) To *reduce* the engine speed to a much lower value at the wheels, so as to achieve reasonable ground speeds.
(3) To provide *variation* of reduction to enable both high-speed and low-speed operations to take place.

These three functions are the reason why transmission systems in cars, trucks, and (particularly) tractors are so complex and expensive. The main reason for needing the system, we can remind ourselves, is due to the mismatch in speeds between the output from the engine and the required input to the wheels. Since the high-speed internal combustion engine is here to stay, we are obliged to accept the mismatch and design transmission systems in order to cope. Before looking at the types of components which can be used in the transmission system, however, another important point must be discussed.

5.2.1 Torque multiplication

We know that power = torque × rotational speed. Given a constant power, what is the effect on *torque* if we take the previous example and reduce the rotational speed, by means of reductions in the transmission system, to only 1/188 of the engine speed? The effect can be seen by examining the equation above. The torque will be increased in the same proportion, to 188 times engine torque. This torque, or twisting force, has to be transmitted by the components of the transmission system, so that the input shaft to the wheel then has, in theory, to be 188 times as strong as the output shaft from the engine. (Intermediate components must transmit intermediate values of torque, dependent on the speed at which they rotate.) In practice it is unlikely that such high torque could be transmitted by the wheels. The reason is that the available engine power from a tractor of this size would be about 50 kW. To use full power at 3.3 km/h (0.91 m/s) would require a pull of 50/0.91 or 55 kN. To produce this pull in average soil conditions the tractor would then have to weigh about 8.6 tonnes. A practical tractor weight is about 3 tonnes (enabling a pull of about 19 kN to be exerted). The torque able to be transmitted by the wheels is therefore likely to be only about 65 times engine torque rather than 188 times, so the final transmission may be designed with some confidence to cater only for the lower torque value. (Although if the wheels *could* get enough traction it would result in transmission failure. The experience of drive line failures in North America, where some farmers habitually ballast their tractors to twice the design values in order to get high traction at low speeds, supports this view.)

Having pointed at the *problems* of torque multiplication it must also be stressed that at low speed it is *necessary* to transmit high torque in order for the power to be produced for traction. For other applications, such as when an engine is used to power stationary equipment (processing plant, pumps, or generators) the input speeds will be very much higher than for traction. The 'transmission system' in this case then often becomes a question of simple connection or perhaps a straightforward reduction of low magnitude.

5.2.2 Transmission components

Most transmission systems are made up of a *series* of components. In the case of a conventional motor truck, for example, the system consists of clutch, gearbox, propellor shaft, differential, and half shafts (in that order). Similar remarks apply to a conventional tractor, except that a propellor shaft is not normally found, and the system may incorporate additional final reductions after the differential.

All these components serve different functions (in line with the requirements of a transmission system outlined above). A clutch allows disconnection, that is, a break in the system; a propellor shaft and/or half shaft provides physical connection between points at a distance, without any change in speed; a differential and final reductions (where fitted) provide fixed reductions in speed; a gearbox provides a series of alternative, selectable reductions.

As it is not easy to obtain a reduction, in one step, of more than about 6:1 it is usually necessary to provide a series of reductions, located at various places along the transmission line (from engine to wheels), and taking advantage of the very useful fact that 'reductions in series multiply up'. Thus a series of three steps in series, for example, each giving a reduction of 6:1, would provide a total reduction of $6 \times 6 \times 6 = 216:1$ (more than enough for the tractor example examined earlier). Such reductions could take place in the gearbox, rear axle, and final reduction of a tractor transmission system.

Similar principles apply to smaller machines, where the components used in the various reduction stages are more likely to be a combination of belts, chains, and gears. The total reduction required can still be considerable—for example a small tractor with drive wheels 0.8 m in diameter and powered by an engine running at 3600 revs/min requires a total reduction in bottom gear of 150:1 to operate at a ground speed of 1 m/s (3.6 km/h). An example of how this can be achieved is:

5:1 belt drive from engine to gearbox;
5:1 reduction in the gearbox (1st gear);
4:1 reduction at the rear axle/differential;
1.5:1 reduction at final 'drop' chain drives;
Total 150:1.

If the gearbox were a three-speed type, giving second and third gear reductions of (say) 2.5:1 and 1:1 respectively, the maximum tractor speeds in the gears would be:

1 m/s (3.6 km/h) in first gear (for ploughing);
2 m/s (7.2 km/h) in second gear (for weeding);
5 m/s (18 km/h) in third gear (for transport).

It can be seen, therefore, that building up a transmission system for a small machine in a series of steps is perfectly feasible, provided that the torque multiplication factor is not overlooked. In this case such a tractor weighing 1 tonne would pull a maximum of about 6.4 kN in bottom gear (due to traction limitations), and at 15 per cent slip would travel at $1 \times .85 = 0.85$ m/s. It could therefore transmit a maximum of $6.4 \times 0.85 = 5.4$ kW of drawbar power. This would require an engine power of about 8 kW (allowing for a gross tractive efficiency of 68 per cent) but the torque at the wheels in bottom gear would still be engine torque × 150 (since the major losses are in slip and rolling resistance). This is quite a high value, and would amount to a torque of about 3150 Nm at the axle. This, however, is the maximum torque which is likely to be transmitted by the wheels, regardless of whether a higher powered engine is fitted. The above example will be used in the more detailed description of transmission component types, to follow.

5.2.3 Types of transmission components

(i) Belt drives. Flat belts, once widely used, now have few applications in mobile machines. V belts, driving on the sides of V-shaped pulleys, have much higher ratings and are more compact, and will be discussed here. As V belts drive by friction alone it is the frictional force (equivalent to μN) which dictates the ability to transmit power (Figure 5.6). If the belt and pulley are in good condition, (μ satisfactory) then the force N applied by the belt to the pulley determines the grip. This force is achieved by ensuring that the belt has sufficient tension.

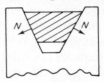

FIGURE 5.6 Section
of V-belt and
pulley

As belts stretch slightly, particularly when new, there must be a way of maintaining the tension, either by varying the centre distance between the two pulleys (input and output) or by using a tensioner. Given that the tension is satisfactory, the ratio of output to input speeds (and the increase of the torque ratio) is given by the ratio of the pulley pitch circle diameter (i.e. the effective diameters at which the centre of the belt runs). Thus with an input pulley diameter of say 100 mm, the nominal output speed with a 200 mm diameter output pulley will be 100/200 or half the input speed (and the torque will be twice as much) (see Figure 5.7).

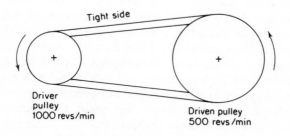

Driver pulley 1000 revs/min

Driven pulley 500 revs/min

FIGURE 5.7 Simple V-belt system

The term 'nominal' is used above to draw attention to the fact that, as V belts are flexible and rely on a friction grip to transmit the power, the actual speed ratio will vary slightly with load, due to creep and (possibly) slip effects. V belts do not give a positive, synchronous drive in the way that chains and gears do; for most transmission applications, however, a slight variation in speed ratio is not important.

Knowing the pulley diameters and the approximate desired centre distance, the nominal belt length can be obtained from simple geometry, and is given by the formula:

$$L = 2C + \frac{(D-d)^2}{4C} + 1.57\,(D+d)$$

where L = belt length, C is centre distance, D is large pulley diameter, d is small pulley diameter. Belts are usually manufactured in a range of sizes so it is necessary to select the belt length nearest to the calculated value of L, and then adjust C accordingly.

For example, if two pulleys of pitch diameters 200 mm and 100 mm respectively are to be positioned 250 mm apart, the formula gives

$$L = (2 \times 250) + \frac{(200-100)^2}{4 \times 250} + 1.57\,(200+100)$$

$$= 981 \text{ mm}$$

Standard belt lengths may be 950 or 1000 mm, so that the centre distance C would need to be adjusted to 234 mm or 260 mm respectively. (Note that a relatively new type of belt, the timing belt, is now used extensively as a camshaft drive on automotive engines, and does provide synchronous drive due to its method of transmitting power to notched pulleys via teeth on the underside of the belt. In an application where it is imperative that the two pulleys stay in exactly the same radial position with respect to one another, these belts should be considered.)

V belts are useful for transmitting power at high speed and low torque. They can easily run at engine speeds (3000–5000 revs/min), do not require lubrication, can tolerate a certain amount of misalignment and a fairly hostile environment, are quite cheap but do not last particularly long in some applications. As mentioned earlier, they may require a tensioner, fitted on the slack side of the belt, to maintain the correct tension—a spring tensioner does this automatically. An advantage arising from this is that if the tensioner has an overriding control to pull it away from the belt (assuming it is fitted on the outside as in Figure 5.8) the belt will become slack and the drive will disengage. The tensioner can thus be used as a clutch, and indeed quite a number of small, belt-driven machines such as single-axle tractors use this principle. The important point is that, to prevent intermittent snatch of the drive when disengaged, the slack belt should rest against rods or bars provided in the positions shown at A in Figure 5.8. The take up is then reasonably smooth, although the reverse bending occurring around the idler pulley will reduce belt life. (This arrangement cannot be used with the high-strength 'SP series' belts.)

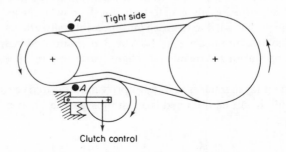

FIGURE 5.8 V-belt with tensioner/clutch

For application to the small tractor described earlier a reduction of 5:1 is required. Pulley diameters of 80 and 400 mm will provide this reduction (although the large pulley is rather large). Transmissable power per belt can be obtained from manufacturers' catalogues. In this case, choosing 'A' section belts running from the engine at 3600 revs/min gives a power per belt of 2.3 kW. Four belts would therefore transmit 9.2 kW, which is in excess of the 8 kW engine power and therefore satisfactory. It should be noted that V belts must always be *over*designed, or early failure will result. (For example, three belts, transmitting a total of 6.9 kW, would not be satisfactory.)

A somewhat different, but useful, type of belt has a notched underside but runs on smooth pulleys. These belts are used in variable V belt drives, fitted to some small machines such as mowers, garden tractors, and all-terrain vehicles. This arrangement can automatically provide a change in ratio from about 3:1 to 1:1 by changing the effective diameters of the (split) pulleys according to engine speed and load (Figure 5.9). They may also incorporate an automatic clutch, and are therefore worth considering where a simple, variable speed drive is required. It is possible that, for sustained high-power operation in difficult conditions, the life of the belts and aluminium pulleys would be suspect but more assessment is needed with these drives at the smallholder level.

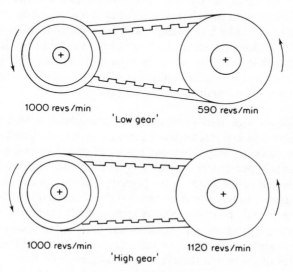

1000 revs/min   590 revs/min
'Low gear'

1000 revs/min   1120 revs/min
'High gear'

FIGURE 5.9 Variable V-belt system

(ii) Chain drives. The characteristics of chain drives are in many ways the opposite of those for belts. Chains are high-strength, precision components requiring accurate alignment, good environment protection, and lubrication. They are more suitable for low-speed, high-torque applications and give a positive drive by virtue of toothed sprockets acting on rollers in each chain link. Correct tension is important but not as vital as with belts. The speed ratio obtained is equal to the ratio of the sprocket tooth numbers—for example, a 19-tooth and a 57-tooth sprocket will give a speed ratio of $57/19 = 3:1$.

Given a good arrangement, lubrication, and protection, chains are designed to last for 26,000 hours, which is much more than the design life of many small-scale agricultural machines. The factor of safety on strength is usually about 10:1. This means that, where a lower design life of say 5000 hours is required it is perfectly possible to *under*design a chain drive by a factor of two, that is to put twice the rated load through it. This is very useful in terms of reducing size and costs. As an example, the small tractor transmission examined earlier could use chain drives as a final reduction to the wheels. This has an advantage in that the axle is able to be located much higher, improving ground clearance over crops.

The reductions suggested for the transmission system were 5:1 in the belts, and a further 5:1 in a gearbox, followed by 4:1 in the differential. At the end of the rear axle the total reduction would therefore be $5 \times 5 \times 4 = 100:1$, giving a speed of $3600/100 = 36$ revs/min as the input speed to the upper sprockets. Using a 15-tooth and a 23-tooth sprocket to give a speed ratio of 1.53:1 (the nearest available to 1.5:1) a chain catalogue gives a rating of 1.4 kW with a 1 inch pitch chain at 36 revs/min. It will be remembered that the chains are likely to be called on to transmit 8 kW maximum (but only for a fairly small proportion of the operating life). However, we can safely double the chain rating to 2.8 kW for a 5000 hour life. Also, with two chains (one for each wheel) the load will effectively be shared—certainly while ploughing in a straight line. The final value obtained is now 5.6 kW. Bearing in mind that the factor of safety on strength is still about 5:1, it is possible that this drive would be regarded as acceptable to transmit the 8 kW. If not, duplex (or double) chain could be specified, giving an increased rating factor of 1.7, and making the rated power 9.52 kW.

(iii) Gears. Gears are used extensively in transmission systems for larger machines. A conventional tractor has a gearbox, differential, and final reduction to provide the necessary step down and choice of ratios, and all will be based on gears of various types (spur, bevel, or epicyclic). The application to smaller machines is less universal, unless a proprietary gearbox or axle is incorporated, since it is not as easy to buy suitable gears 'off the shelf' as it is to obtain belts and pulleys, or chains and sprockets. Nevertheless, simple gears may be made fairly easily, given certain production equipment and material, and selection gearboxes provide a very convenient means of varying the ratios for different tasks. The basic principles of gears will therefore be briefly covered here.

Like chains, gears give a positive drive, the speed ratio being equal to the ratio of the number of teeth. They also need accurate alignment, lubrication, and environment protection. If well designed they should last for an equivalent number of hours (e.g. 26,000) but can in fact be designed to last for any given design life. The main differences in operation, compared with chains, is that the centre distance is less (equal to the sum of the pitch circle radii of the two gears) and the gears rotate in opposite directions. (Inserting an 'idler' will change this, without affecting the speed ratio.)

Gear trains is the term given to a number of gears in series. These trains may be simple, or compound (more than one gear on a shaft), or epicyclic. Simplified diagrams show these types (Figure 5.10). Gears which allow the drive to turn through a right angle are known as bevel gears; the teeth are formed on a cone rather than a cylinder. Most simple gearboxes use spur gears (teeth cut straight) or helical gears (teeth cut at an angle). Gears can be engaged and disengaged either by sliding them into engagement or by the use of dog clutches (or synchromesh cones in some cases).

Most differentials use a pair of bevel gears (crown wheel and pinion) to get a right-angle turn in the drive in addition to a reduction (of around 4 or 5:1), followed by what is effectively an epicyclic gear set which enables the two half

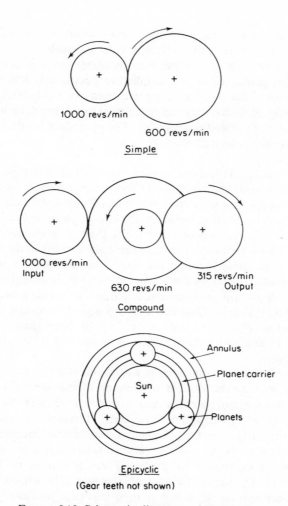

FIGURE 5.10 Schematic diagrams of gear trains—
simple, compound, and epicyclic

shafts to rotate at different speeds for turning corners. (This facility can be a problem when pulling hard in a straight line, as the drive wheel with the lowest grip can start to spin. Most tractors are therefore fitted with a differential lock which in effect locks out the epicyclic section, forcing the half shafts to rotate at equal speeds.) Some large tractors are fitted with conventional epicyclic reductions positioned in the drive line after the differential, that is one on each half shaft, in order to obtain the total reduction required. Some small tractors (and occasionally large ones) may have spur gear reductions close to the drive wheels in order to increase the ground clearance under the rear axle.

It can be seen therefore, that a number of small machines will be fitted with some kind of gearbox in order to obtain a choice of ratios, and will also usually have a differential (if there is more than one drive wheel).

To design a transmission system incorporating gears is not particularly difficult, but as the gears need to be positioned very accurately with respect to one another, the design of the gearbox casing (either cast or welded) is critical. Gears are normally designed so that they will either wear out or break up (due to fatigue effects) at the end of their useful design life. The procedure for designing gears is laid down in standards such as BS 436.

(iv) Hydrostatic drive. This name implies the transmission of power using the pressure of a fluid (rather than its velocity, as in the hydrodynamic drives used in 'automatic' motor car torque converters). In fact the fluid does move around the circuit. The basic principle is to use an engine-driven pump to displace hydraulic fluid from the low-pressure side of a circuit into the high-pressure side, at the end of which is a hydraulic motor (or motors) connected directly to the wheels. As the displaced fluid must pass through the motor, it causes it to rotate thereby transmitting power to overcome the load on the wheels. It should be noted that it is the load which causes the pressure in the fluid to rise, so that power transmitted equals pressure × flow. If the load becomes too high the pressure will rise above the design value (usually around 300 bar, i.e. $30 \times 10^6 \text{ N/m}^2$) and relief valves will blow, allowing the fluid to return to the low-pressure side of the circuit, normally via an oil reservoir.

This may appear a rather cumbersome way of transmitting power from engine to wheels and in fact the double conversion (from mechanical energy to hydrostatic and back again) normally results in a lower efficiency than mechanical drives (perhaps 70 per cent instead of 90 per cent). However, there are a number of advantages, particularly when the pump has a controllable variable displacement (from zero to maximum). At zero displacement, equivalent to 'neutral', no oil flows and therefore the motors (and wheels) remain stationary. Increasing the pump displacement by means of a control lever starts the wheels rotating, their speed being dependent on the pump setting. It is thus possible to increase the tractor or vehicle speed to any value up to maximum using one control lever. No clutch, gearbox, or even brakes are required.

By specifying large-capacity wheel motors (sometimes with built-in epicyclic final reductions) the low ground speeds required by tractors can be accommodated. Because the only connection between engine and wheels is by means of flexible hydraulic pipes, they may be mounted in any convenient position relative to one another without the need for a complex transmission system to connect the two. These overall advantages are considerable (particularly the one-lever control system which provides extremely simple operation).

Why, then, is the hydrostatic system not more widely used? In fact, at least one small tractor designed for smallholder conditions uses this principle. The main drawbacks are probably in the areas of cost and maintenance. Hydraulic pumps, motors, valves, etc. are precision components, and rather expensive as a result. Because they use such small clearances between their working parts it is essential that absolute cleanliness of the hydraulic oil is maintained. This can be achieved in operation with the aid of full flow-filter systems, but repair or replacement of components in the field calls for extreme care to avoid contamination.

The application of high-precision systems such as hydrostatics raises an issue of whether machinery for smallholder agriculture should be high-technology and tamperproof, or of a much simpler—even cruder—form such as spur gears, belts, and chains, so that breakdowns offer at least the possibility of running repairs using village-mechanic level skills. This issue is often raised in discussions centred on the choice of technology for the smallholder situation, but is not really resolved as yet.

REFERENCES

Bacon, D. M., Bacon, N., Moncrieff, I. D., and Walker, K. L. (1982). *The Effects of Biomass Fuels on Diesel Engine Combustion Performance* (Perkins Engines Ltd.).

Matthews, J. (1982). Effective choice and use of agricultural tractors, *The Agricultural Engineer* (UK), **37**, 3, 96–9.

Stevens, G. N. (1982). Equipment testing and evaluation (NIAE, Silsoe.).

6 *Traction, Steering, and Braking*

Traction is
dependent upon soil
conditions, wheel
loading and tyre
size—here a small
tractor is being
evaluated in Central
Africa

6.1 INTRODUCTION

The primary purpose of a wheel is to support a load whilst moving with minimum resistance over a surface. In addition the wheel may be required to produce, at its contact with the ground, forces to provide tractive, braking or steering action (Figure 6.1).

The first practical pneumatic-tyred wheel was developed in 1888 by J. B. Dunlop. He fitted his tyre to a bicycle wheel, providing greatly increased comfort for the rider due to the tyre's shock-absorbing qualities, in addition to the essential functions stated above. Further development was necessary before pneumatic tyres suitable for trucks could be produced in the early 1920s.

Operating conditions in agriculture are considerably more rigorous than those for road transport and it was not until the early 1930s that tyres with an adequate field performance were generally available. The additional problems to be overcome arose from the deformable and variable nature of the soils on which agricultural equipment normally operates, from the rough surface conditions

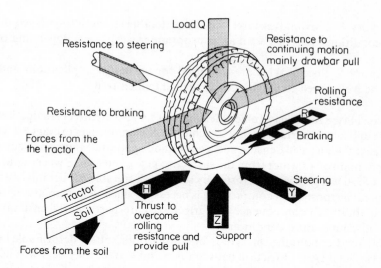

FIGURE 6.1 Forces acting on a wheel. Forces from the vehicle are
applied to the wheel through the hub and hub bearings. Forces from
the soil are applied to the wheel through the tyre/soil contact

and from the high pull which an agricultural tractor is commonly asked to
provide for heavy cultivation machinery.

The object of this chapter is to bring together practical and theoretical aspects
of agricultural tyre performance so that the tyre user may have a good
understanding of problems involved and be able to use his equipment most
effectively.

6.2 TYPES OF TYRE USED IN AGRICULTURE

Manufacturers provide a wide variety of tyre types, each intended to suit a
particular set of functional requirements. Each tyre type is usually available in a
range of sizes to allow the capacity of the tyre to be matched to the size and power
of the equipment. The functional requirements of the tyre influence the tread
pattern provided and this feature gives a good visual guide to intended use, the
main categories of which are given below. In the USA a uniform code for tyre
categories has been initiated by the Tyre and Rim Association and adopted by
the American Society of Agricultural Engineers.

Category 1: Undriven steered wheels (front tractor) usually have deep
circumferential ribs to assist steering action in soft soils.

Category 2: Driven wheels (rear tractor) have tread bars of various heights,
angles, and patterns for particular, or general purposes.

Category 3: Implement tyres usually have considerable section width for
maximum load-carrying capacity with minimum diameter to provide low
trailer platform height.

Category 4: Earthmover tyres need considerable strength for rugged operating conditions, together with an appropriate tread pattern depending on soil type.

Category 5: Special tyres are produced for particular applications and may not be well suited to work outside their intended role.

A typical tyre for a road vehicle depends for its grip on direct friction between the tyre and a hard surface so that to achieve a reasonable performance it has low tread height, high tread density, and a continuous pattern. Many small cuts assist the rapid removal of water from the surface to give good grip in wet conditions at speed. Good wet grip is also assisted by selecting a suitable rubber mix, although for general purpose use a compromise is necessary since wear increases with wet-grip capability. In consequence wet-grip mixes are particularly unsuitable for dry, high-temperature conditions.

In off-road conditions the tyre is operating on a softer, deformable surface where the coefficient of friction between tyre and surface may be low. The tread depth is increased and the tread density decreased giving distinct lugs, with spaces between, which enable the tread to penetrate and grip the soil. The thrust which the tyre can produce is more dependent on the strength of the soil in shear and less on tyre-to-soil friction. Surface conditions, such as stones and flints, require the rubber mix to be selected for resistance to tearing.

As the tread depth increases stiffness in flexing is reduced and tread movement assists self-cleaning in wet soil conditions but wear on hard surfaces becomes a problem. Soft-soil tyres are liable to more rapid wear on road surfaces and the separate lugs also give rise to problems of noise and vibration. Tyres having a deep tread—about 50 mm compared to about 35 mm normally—are available and may be used on loose deep soils. Increased tread density provides a more continuous running surface, but at the expense of self-cleaning ability; it will be of value in reducing wear when a considerable proportion of hard surface running is undertaken and absolute maximum off-highway tractive ability is not required. Trailer tyres are designed to give low rolling resistance. The tread depth is shallow and the continuous circumferential ribs give smooth running. For those tyres intended for soft soil the tread is designed to give safe trailing characteristics on corners. Implement driving tyres have a pronounced 'tractive' tread pattern to transmit drive from the soil to the implement mechanism.

Steered wheels usually have plain circumference ribs to give high resistance to side forces and for softer soils the ribs are made progressively taller and thinner. There may be a tread pattern to improve control by encouraging the wheel to maintain rotation, particularly with tractors using mounted implements when weight transfer reduces the load on the front wheels.

Flotation tyres have very large sections compared to their diameter, and work at low inflation pressures. These factors combine to give a very low ground pressure for soft conditions. The tread depth is shallow and tread density high, giving reasonable pull with minimum damage to the ground surface.

Industrial tyres are usually designed to work on hard surfaces and so tend to have treads similar to road tyres. They often have special rubber mixes to resist chemical attack, for example from industrial oils. They normally work at much higher pressures (combined with small physical size) for maximum load-carrying capacity on hard surfaces. Tyres used on fork lift trucks may be semi-pneumatic, having a foam filling—cushion tyres—as there is then less danger of shedding the load on an extended mast if the tyre is punctured. Solid tyres also overcome this problem but give ride problems causing more severe loading on the machine and operator.

6.3 TYRE CONSTRUCTION AND MATERIALS

6.3.1 Construction

A pneumatic tyre consists of a flexible casing built of textile cords laid in specific directions to restrain extension of the casing on inflation with a gas—usually air—and to carry the consequent inflation stresses. The cords are held in place by a matrix of rubber which holds them apart and allows a certain movement between the cords by shear deflections. The cord plies pass round the bead wires which form two hoops of high-tensile steel wires which resist the outward tensions caused by inflation. The modulus of the various materials varies greatly, and, because the geometry of the construction makes the tyre a highly redundant structure, it is very difficult to analyse the stresses theoretically. Consequently most work on tyre construction is of a semi-empirical nature. The tread rubber is bonded to the casing and is of a form and mix type to suit the operating conditions (Figure 6.2).

The cord layers (plies) can be laid in a variety of ways, considerably affecting operational characteristics (Figure 6.3). Cross-ply is the normal construction for agricultural tyres. Both side walls and tread are substantially rigid to longitudinal and side loads; the degree of rigidity can be changed by altering the angle of the

Radial ply

Cross ply

FIGURE 6.2 Ply construction of tyres

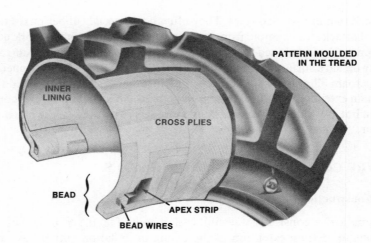

FIGURE 6.3 Cross-section of a tractor rear tyre

plies, measured from the plane of the diameter. Radial ply construction gives relatively flexible side walls together with a very well-braced tread, giving longer tread wear lift on hard surfaces and a smaller drift angle for a given side load.

For tyres working under hard surface conditions where there is risk of tyre penetration, for example in rock quarries, one of the belted layers is often made of shredded wire to provide a steel breaker. To maintain inflation pressure the tyre may be fitted with a separate air container (tube) or be constructed to operate without a tube, in which case the tyre has an integral liner made of a rubber mix, usually a butyl type, which is impervious to air. A tubed tyre will need to have a flap fitted to protect the tube from damage by contact with the rim. Tubeless tyres have the advantage of easier fitting and maintenance, do not suffer from torn out valves if the tyre slips on the rim due to momentary overload, and deflate more slowly under minor punctures.

6.3.2 Materials

(i) Textiles. Originally cotton was used as ply material but during the past twenty years manmade fibres have been used exclusively. Rayon is extensively used but is now being superseded by nylon which has a much higher strength per cord. It must be carefully processed by hot setting to keep the elongation within the required limits. Polyester has bonding and processing problems, does not have the moisture absorption problem of nylon, but has better dimensional stability. It also gives low growth of the tyre in service.

Glass fibre is suitable only for those areas which are not required to flex much, thus it may be used in the belt region but not for side walls. Steel may be used for highly rated tyres, when one ply layer can replace several of textile. It is not used for tractor tyres which have a relatively low rating, but is well suited for use in industrial tyres.

(ii) Rubber mix. The first tractor tyres were introduced when natural rubber was widely used in tyre construction. Most of the rubbers used today are synthetic. Tractor tyres are resistant to most chemicals likely to be encountered with the exception of diesel oil which causes rapid degradation of tread rubbers. Earthmover tyres are usually made with a high percentage of natural rubber because they have thick sections which are liable to excessive hysteresis heating, arising from flexing, if made with synthetics.

6.4 TYRE RATING AND CAUSES OF FAILURE

The load-carrying capacity of a tyre depends on a number of factors: tyre size; construction; inflation pressure; operating speed; duration of load.

6.4.1 Tyre size

Basically the method of specifying tyre size has been to quote the section width and nominal rim diameter, each being specified in inches (Figures 6.4 and 6.5). Thus $11-28$ indicated a section width b_c of 11 inches and rim diameter d_r of 28 inches resulting in an approximate overall tyre diameter of $d_r + 2h = 50$ inches, because in early tyres $b_c = h$. (For an explanation of the symbols used see the appendix to this chapter.) About 1955 extra-wide base rims were introduced for rear tractor tyres, resulting in a wider cross-section for the same section height and hence a lowering of the aspect ratio h/b_c from about 1.0 to about 0.85. The new section width was included with the old or nominal section width in the tyre, designated thus:

$$12.4/11-28$$

The old section width is now being dropped from the size designation so that the tyre quoted will be known simply as $12.4-28$. Recently tyres with even lower aspect ratios have been introduced and this is indicated by including the figure in

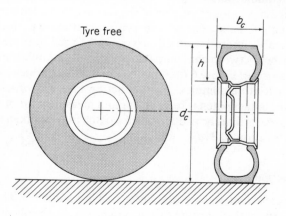

Tyre free

Figure 6.4 Tyre free

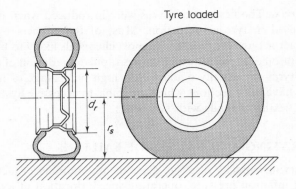

FIGURE 6.5 Tyre loaded: d_c—overall diameter (catalogue); b_c—overall width (catalogue); r_s—static loaded radius; d_r—rim diameter; h—section height; $\dfrac{h}{b_c}$—aspect ratio

the tyre designation thus:

$$9.0/75-18$$

This example indicates a tyre with section width of 9.0 inches and an aspect ratio of 0.75 (75 per cent), used on an 18 inch diameter rim.

The static loaded radius of a tyre is usually included in manufacturers' tyre data tables, giving a value for maximum pressure and load when standing on a hard surface. This can be taken as the rolling radius under such conditions and the distance moved in one revolution of the wheel will be $2\pi r_s$, assuming no slip. When the wheel is operated on a soft surface, which may deflect more than the tyre, the rolling radius is not readily measured directly. It is derived by measuring the distance moved by the wheel in one revolution without slip and calculating the equivalent rolling radius.

The area and dimensions of the part of the tyre in contact with the ground, that is the 'contact patch', have a considerable effect on tractive performance and are thus of considerable interest. Lattice curves can be used to present information on the relationship between contact patch area: load and inflation pressure for a tyre on a hard surface (Figure 6.6). For a tyre on soft soil, which also deflects, the situation is much more complex. For a traction tyre it is often convenient to assume that the contact patch is rectangular and measurement of a large number of such tyres suggests that the width of such a patch may be taken as about $0.87 \, b_c$ and its length as about $0.31 d_c$.

6.4.2 Construction

Details of tyre construction have been discussed in Section 6.3. The load capacity depends on the strength of the tyre casing as indicated by the ply rating of the tyre. Originally the ply rating specified the number of layers of cotton used to give

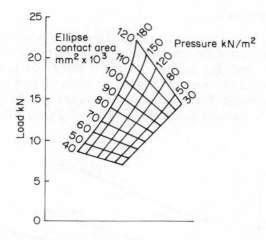

FIGURE 6.6 Area of contact (ellipse) versus load
and inflation pressure for 14.9–28 (6-ply rated
tyre)

tensile strength to the casing. Other materials with a higher tensile strength than
cotton are now used in tyre construction and the ply rating is used only as an
index of tyre strength and does not necessarily state the actual number of plies.
The strength of automotive and truck tyres is now specified by quoting a load
range rather than a ply rating (Figure 6.7).

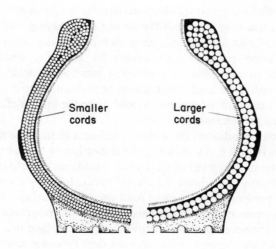

FIGURE 6.7 Alternative constructions for tyres of
equivalent ply ratings

6.4.3 Inflation Pressure

When carrying a load over a hard surface the tyre deflects, causing increase of the contact patch area, until the inflation pressure acting on the contact patch area can support the load. For agricultural tractor and trailer tyres the maximum deflection is limited to about 18–20 per cent of the section height in order to prevent casing damage. With increasing load on the tyre the inflation pressure must be raised to maintain acceptable deflections (Figure 6.8).

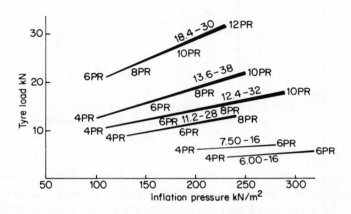

FIGURE 6.8 The relationship between load, inflation pressure, and ply rating

The load–deflection–pressure characteristic for a tyre may be plotted graphically in lattice form (Figure 6.9). A family of curves having load as ordinate and deflection as abscissa are plotted, each member of the family representing a particular inflation pressure and having its zero for deflection displaced proportionately to the inflation pressure. Lines for a particular deflection at various combinations of load and inflation pressure may be drawn in and, when extrapolated to zero inflation pressure, will show the load–deflection properties of the casing to be determined.

In soft off-road conditions the soil also deforms to increase the contact patch area (Figures 6.10 and 6.11), reducing the deflection of the tyre for a given load and inflation pressure. Hence in soft ground conditions the inflation pressure may be reduced slightly by about 15kN/m^2 or (2lbf/in^2) compared with the recommended pressure for the same load on a hard surface. This may help to reduce rolling resistance and improve traction in poor conditions but the recommended pressure must be restored for hard surface use.

When tractors are shipped or transported their tyres are commonly inflated to 200kN/m^2 (30lbf/in^2) to prevent them bouncing in transit. They must be set at their correct working pressures before use.

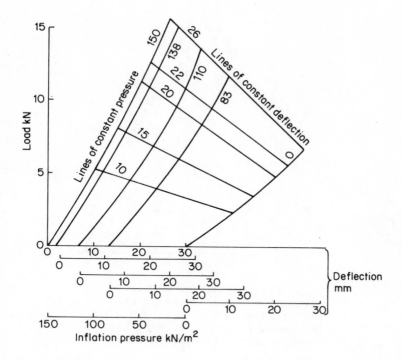

FIGURE 6.9 Lattice plot of load deflection/pressure characteristics for a
12.4–36 tractor tyre

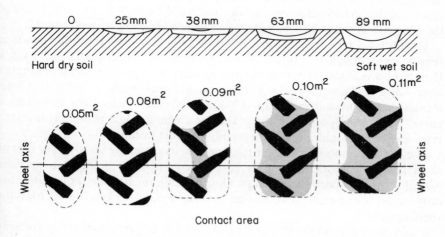

FIGURE 6.10 Contract area between tyre and soil for different soil conditions. Tyre
9–40, 690 kg, 83 kN/m² inflation pressure

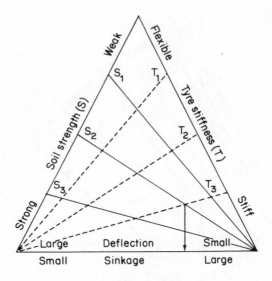

FIGURE 6.11 Schematic representation of the relationship between soil strength, tyre stiffness, tyre deflection and sinkage (after Soane, Dickson and Campbell, 1982. Reproduced by permission of Elsevier Scientific Publishing Co.)

6.4.4 Operating speed and duration of load

A loaded tyre is deflected near the contact patch and is being continuously flexed as it rolls along. The mechanical energy required to flex the tyre is not fully recovered due to hysteresis loss causing heating of the rubber. The fatigue life of the ply materials may also be reduced by repeated flexing at high speeds.

Tractor tyres are normally rated at 30 km/h (20 mile/h) and seldom operate at speeds high enough to cause heat degradation. At lower speeds, say 8 km/h (5 mile/h), a rear tractor tyre is allowed to carry a 50 per cent increase in rated load. For a rear tractor working with mounted implements a 25 per cent load increase may be used at speeds up to 20 km/h (12 mile/h) since for a considerable part of the time only a proportion of the implement's weight is supported on the tyres and the maximum loads are mainly of short duration.

The load on combine harvester tyres may be increased by up to 40 per cent of rated load in conjunction with a 25 per cent increase of inflation pressure provided that the speed is kept below 8 km/h (5 mile/h). Because the combine has only a short operational life compared to a tractor a shorter tyre life will meet the requirements.

On front tyres a 100 per cent increase in load may be permitted for tyres of 6 ply rating and above, at speeds up to 8 km/h (5 mile/h) when maximum loads are of short duration.

6.4.5 Tyre failures

Most tyre failures in agriculture are due to excessive casing deflection leading to breakdown of the casing plies caused by overflexing. Traction tyres may also carry a high tangential load, often nearly as great as the vertical load, causing tangential deflections leading to excessive deformation of the tyre if combined with too high a vertical deflection. At low pressures the tyre may creep on the rim, causing tube failure—the tyre bead seat on the rim can have cross-grousers to prevent this on special low pressure tyres. Figure 6.12 illustrates correct tyre inflation.

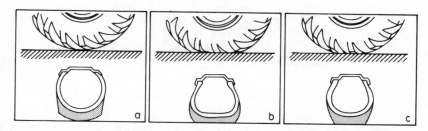

FIGURE 6.12 Tyre inflation: (a) overinflation; (b) underinflation; (c) correct inflation

Very rapid wear may result from the tyre rubbing on the chassis or fenders, or on a drawbar when on maximum turn and such situations should obviously be avoided. Tyres can be relugged provided the casing is not structurally damaged.

6.4.6 Tyre bursts

When new the bursting pressure for a tyre is about ten times the operating pressure, but after use this can be very considerably reduced by defects which may not be visible.

Tyre failure by bursting has often led to fatal consequences and great care must be taken when fitting and inflating tyres. An inflation cage to protect the operator is essential (Figure 6.13). Calculation shows that a 13.6–38 tyre inflated to a pressure of $100 \, \text{kN/m}^2$ (14 lbf/in^2) has sufficient energy to accelerate a 65 kg (10 stone) man to a speed of 100 km/h (65 mile/h).

6.4.7 Tyre removing and fitting

When removing and fitting tyres great care must be used to avoid damage to the tyre and to guard against bursts, with their possible disastrous consequences. Removing and fitting can be undertaken either with the wheel on the tractor or with the wheel off. With the wheel on the tractor it is essential for adequate safety that the tractor is standing on firm ground, the hand brake is working effectively and firmly applied, and axle stands are used, seated on a firm base.

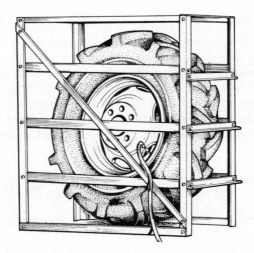

FIGURE 6.13 A tyre-inflation cage

Inextensible wires are built into the beads of tractor tyres. Attempting to stretch them over the rim flange is useless and may result in damage to the beads. Removal and fitting can be made easier by using an approved tyre bead lubricant applied to the beads and to the tyre levers and by ensuring that the beads are carefully adjusted into the well of the rim opposite the point of leverage.

Before fitting tractor drive tyres and implement driving tyres a check should be made that the tread pattern will be in the correct direction when the wheel is fitted to the tractor of implement. If the direction is incorrect self-cleaning and drive will be adversely affected.

When seating the beads of a newly fitted tyre the inflation pressure must *never exceed* $140 \, kN/m^2$ ($35 \, lbf/in^2$). If the beads do not seat correctly the airline must be disconnected, the assembly deflated and the cover and tube relocated on the rim before trying again. The pressure must be adjusted to its recommended level after fitting.

6.4.8 Tyre storage

Tyres and tubes may suffer premature deterioration if stored in unsuitable conditions. Particular attention should be paid to the following points.

(i) Light. Ultra violet rays have a damaging effect so tyres should be protected from sunlight by covering with tarpaulins or closely woven material. Electric lights should be turned off. Tubes should be kept in their boxes or black polythene bags until needed.

(ii) Oxygen and ozone. Tyres age more rapidly when in contact with oxygen and ozone. Excessive circulation of air helps to provide oxygen to support this

process, so draughty conditions should be avoided. Electrical discharges produce ozone, which is particularly harmful in causing cracking, and tyres should therefore not be stored close to electrical equipment.

(iii) Temperature. Temperatures between 5°C and 15°C (40°F and 60°F) are ideal for storage. Direct contact with heat should be avoided.

(iv) Liquids. Water may cause damage by seeping through a small cut into the tyre fabric, consequently water should not be allowed in contact with tyres. Petroleum products, whether in solid, liquid, or vapour form, are readily absorbed by the rubber compound, causing damage, and tyres should be stored away from such products.

6.4.9 Chemical attack by liquids

Various liquids may cause tyre damage of either a permanent or temporary nature. Permanent damage occurs, for example, when petrol or oil is allowed to saturate a tyre, resulting in a spongy distortion of the tread and rapid wear in service. Common materials which cause permanent damage are: engine and gear-box oils, greases, fuel oil, kerosene, hydrogen peroxide, hot tar, white spirit, and most brake fluids. If contamination by these liquids occurs the tyre should be wiped clean as soon as possible and washed with synthetic detergent in warm water to restrict further damage. Tyres should never be left standing in patches of the liquids mentioned above. Some liquids may cause a similar swelling and softening of the rubber, but the rubber will return to its original state when the liquid dries off. Examples of liquids which cause temporary damage are: petrol, benzene, carbon tetrachloride, ethyl acetate, and turpentine.

6.5 PERFORMANCE OF THE AGRICULTURAL TRACTOR

When the tractor is in work the force exerted on it by the implement may be split into three components:

F_x, the longitudinal component (drawbar pull), opposes the forward movement of the tractor;
F_y, the sideways component, when present, gives rise to steering problems;
F_z, the vertical component, can cause a significant extra load on the rear wheels.

As a result the tractor tyres may be called upon to provide traction, steering effect, and extra load-carrying capacity. These force components exist whether the implement is a trailed one, pulled by a drawbar, or a mounted one attached by a hydraulically controlled three-point linkage. F_z, the vertical component, is usually considerably larger for a mounted than for the equivalent trailed implement and the load-carrying capacity of the tractor's rear tyres needs to be greater in consequence.

The drawbar power, P_F, is the product of the drawbar pull and actual forward speed of the tractor.

$$P_F = F_x \times v \qquad 6.1$$

For example a tractor working at 1.6 m/s (3.6 mile/h) against a drawbar pull of 12 kN (2700 lbf) is producing a drawbar power:

$$P_F = 12\,\text{kN} \times 1.6\,\text{m/s} = 19.2\,\text{kW}\ (25.7\,\text{hp}).$$

The actual forward speed v of the tractor is less than the theoretical forward speed v_n as calculated from the rotational speed and rolling radius of the wheel. This means that in a given interval of time the actual forward distance d travelled by the tractor is less than the theoretical distance d_n. The lost forward speed or distance may be compared with the theoretical quantity and this fraction is called the 'slip' or 'travel reduction' s.

$$\text{Slip} = \frac{\text{lost travel distance}}{\text{theoretical travel distance}}$$

$$= \frac{\text{theoretical distance} - \text{actual distance}}{\text{theoretical distance}}$$

i.e.
$$s = \frac{d_n - d}{d_n} = 1 - \frac{d}{d_n} \text{ (in terms of distance)} \qquad 6.2a$$

or
$$s = \frac{v_n - v}{v_n} = 1 - \frac{v}{v_n} \text{ (in terms of velocity)} \qquad 6.2b$$

If, for example, the tractor mentioned above had a theoretical forward velocity of 1.8 m/s then the slip would be:

$$s = 1 - \frac{1.6\,\text{m/s}}{1.8\,\text{m/s}} = 1 - 0.89 = 0.11 \text{ or } 11\%$$

Slip may be used to define the efficiency of conversion of potential forward movement of the tractor into actual forward movement

$$\eta_s = 1 - s \qquad 6.3$$

when η_s may be called the slip efficiency.

Thus the slip efficiency of the above tractor is:

$$\eta_s = 1 - 0.11 = 0.89 \text{ or } 89\%$$

The thrust H, which the driving wheels produce at the ground must be great enough to:

(a) provide traction to meet the required drawbar pull F_x;
(b) overcome the total rolling resistance R at front and rear wheels of the tractor;
(c) overcome the downhill component of the tractor's weight $G_t \sin \beta$, when moving up a slope.

In the discussion following it will be assumed that the tractor is working on level ground. Then:

$$H = F_x + R \qquad\qquad 6.4$$

F_x is the useful fraction of the total thrust produced and a rolling resistance efficiency η_R may be defined:

$$\eta_R = \frac{F_x}{H} = \frac{F_x}{F_x + R} \qquad\qquad 6.5$$

When considering the power transmitted from the engine to the driving wheels the efficiency η_t of the transmission (gearbox, differential gears, etc.) must be taken into account.

If these three efficiency factors are known, or can be estimated, the gross tractive efficiency η_F may be calculated as their product:

$$\eta_F = \eta_s \times \eta_R \times \eta_t \qquad\qquad 6.6$$

This product defines the efficiency of conversion of engine power into drawbar power.

Figure 6.14 illustrates these aspects of performance by indicating the distribution of engine power output as between useful work and lost work. Wide variations in gross tractive efficiency can occur as operating conditions vary. On a good road surface a particular 40 kW tractor can give 33.2 kW Drawbar Power at 23.6 kN pull with 8 per cent slip, and 83 per cent overall efficiency. Whereas the same tractor on a freshly cultivated field will give only 25.2 kW Drawbar Power at 19.5 kN pull with 15 per cent slip, and 63 per cent overall efficiency.

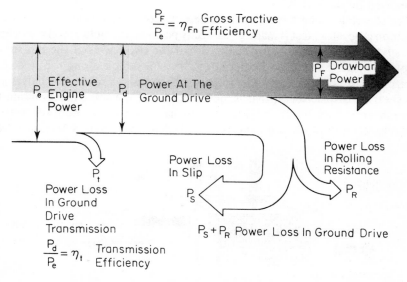

FIGURE 6.14 Power input/output relationship for a tractor ground drive

6.6 PRINCIPLES OF OFF-HIGHWAY MOBILITY

A vehicle operating on the highway travels over a rigid, virtually non-deformable, surface. In contrast, an off-highway vehicle works on soil, a deformable material which yields under load. A vertical load causes vertical yield, or sinkage, which continues until the soil builds up sufficient resistance to support the load. A horizontal load, such as that applied to the soil by a driven wheel in order to generate traction (thrust), will deform the soil horizontally.

The study of off-highway mobility depends on a knowledge of soil strength, in particular the load-deformation relationships for soils when loaded vertically and horizontally. These two modes of deformation interact but for most purposes they may be studied separately.

Agricultural, forestry, earthmoving, and specialized military vehicles commonly operate in off-highway conditions and the discussion which follows may be applied to these classes of vehicle. The discussion will concentrate on wheeled vehicles but much will apply also to tracked equipment.

6.6.1 Soil strength

Soil usually has very little tensile strength. When acted on by a compressive load some reduction of volume—compaction—occurs, but it will support a continual increase in load without failure. The manner of soil failure in a working situation is by shearing. Even though the load applied to the soil may be compressive, such as when a vertical load is applied to a soil surface or an implement tine is pulled through the soil, the effect will be felt on particular planes in the soil as shearing stresses. It is these planes that will limit the maximum load which can be supported by the soil.

The shearing strength of soil may be tested in the field or in the laboratory. A common laboratory test is made by carefully packing a sample of soil into a heavy brass box which is split horizontally so that the two halves of the box may be displaced relative to each other, shearing the soil sample along the area of the horizontal join (Figure 6.15). The relative displacement j of the box halves gives the shearing displacement, that is, deformation, of the soil corresponding to a given shearing load. It is found (Figure 6.16) that for a particular normal stress the shearing strength of the soil increases as it is deformed, up to a maximum value

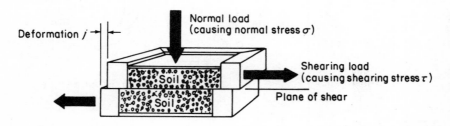

FIGURE 6.15 Soil shearing test using translational shear box

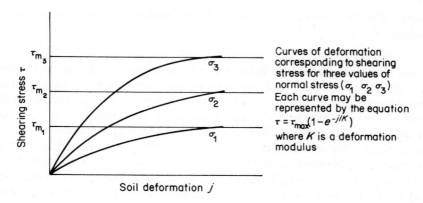

FIGURE 6.16 Soil shearing strength characteristics—curves of deformation

τ_{max}. After this value is reached no additional strength occurs despite further deformation. An exponential equation of the form

$$\tau = \tau_{max}(1 - e^{-j/K})$$ 6.7

is used to represent the shearing stress/deformation curve, K being a deformation modulus which varies depending on how rapidly, with deformation, the soil strength reaches its maximum value. The modulus has a large value for well-compacted soils. It must be stressed that this equation gives only an approximation as to how soil behaves—there is considerable variation depending on the type and state of the soil—but it does provide a good basis for building a theory of traction.

The maximum shearing strength of the soil usually varies depending on the normal stress which is applied across the shearing area. Figure 6.17 shows the

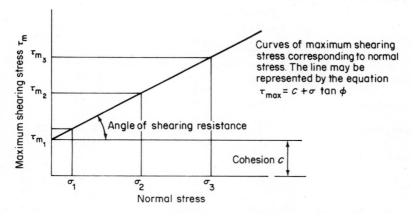

FIGURE 6.17 Soil shearing strength characteristics—curves of maximum shearing stress

manner of this variation, which may be represented by the equation derived by Coulomb in 1773:

$$\tau_{max} = C + \sigma \tan \phi \qquad\qquad 6.8$$

This equation shows that there are two components making up the soil shearing strength. One of these, the 'cohesion' c, is regarded as having a constant value whatever the normal stress. The other component, $\sigma \tan_\phi$, depends on the 'angle of shearing resistance' ϕ, and increases in proportion with increasing normal stress. The values of c and ϕ vary with the type of soil and its condition (Figure 6.18). Clay soils are generally 'cohesive' soils with lower values of ϕ, so that their strength varies less with normal load than sandy soils. Sandy soils are said to be 'frictional', having larger values of ϕ and lower values of c. A loose dry sand has virtually no cohesive strength, though this may build up as moisture content increases.

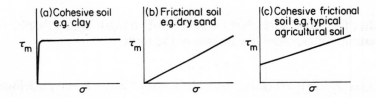

FIGURE 6.18 Shearing strength of various soil types

Agricultural soils vary within these extremes, usually having a certain amount of both cohesive and frictional strength. Typical values are shown in Table 6.1.

TABLE 6.1 Soil Types and Physical Strength

Soil type	Typical particle size mm	State	$\phi°$	C kN/m²
Medium grained sands	1.1	compacted	38–40	–
		loose	32–35	–
Humus sands	.5–.8	compacted	25–30	–
		loose	18–22	–
Loam sands	.02–.2	friable	24–28	20–25
		plastic	24–28	10–15
Loams	.011	friable	22–26	25–30
		plastic	15–19	15–20
Clay	.002	friable	17–19	40–60
		plastic	10–14	25–30

Typical probable values of K (mm) 12 compacted
                                      40 loose

6.6.2 Ideal maximum thrust

The maximum thrust which can be obtained from a wheel or track will depend on the maximum shearing stress which the soil can provide and the area over which this stress is generated. The maximum shearing stress is given by:

$$\tau_{max} = c + \tan \phi \qquad 6.9$$

and multiplying by the contact area S gives

$$S\tau_{max} = Sc + S \tan \phi \qquad 6.10$$

$S\,\tau_{max}$ will be the 'ideal maximum thrust'.
H_{max} which the wheel can produce and the ground pressure under the wheel, will have an average value:

$$\sigma = \frac{Q}{S} \qquad 6.11$$

where Q, the load on the wheel, is divided by the contact area S. Then:

$$H_{max} = Sc + Q \tan \phi \qquad 6.12$$

which is often referred to as the 'Micklethwaite equation'.

It should be noted that this ideal thrust cannot be achieved in practice since it is not possible to deform all of the soil under a wheel sufficiently to provide the maximum shearing stress, even when wheel-slip is 100 per cent.

6.6.3 Thrust/slip relationship

In normal work the soil under the driving wheels is not deformed sufficiently to generate its maximum shearing strength, and the thrust produced is less than the ideal maximum thrust. When a wheel is operating with steady slip s the horizontal soil deformation under the wheel is found to increase steadily from the front to the rear of the contact length between the wheel and the ground (Figure 6.19). For a given contact length and value of slip a deformation pattern will exist which will correspond to a particular thrust H. If this thrust is expressed as a proportion a, called the 'thrust ratio', of H_{max}, that is

$$a = \frac{H}{H_{max}}$$

then a graph (Figure 6.20) may be drawn relating the thrust ratio to a 'slip ratio'

$$J_k = \frac{sl}{K},$$

This graph may be used in conjunction with the Micklethwaite equation to estimate the slip-thrust performance of a tractor, and the slip-pull performance.

Example:
A wheeled tractor has a combined loading on the rear driving wheels of 26 kN

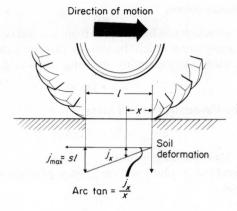

FIGURE 6.19 Soil deformation under a tyre with slip. At a point distance x along the contact length l soil has been deformed through a distance $j_x = sx$

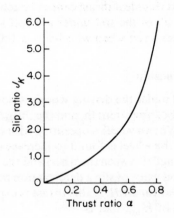

FIGURE 6.20 Thrust-slip function.

$$\text{Thrust ratio } \alpha = \frac{H}{H_{max}}$$

$$\text{Slip ratio } J_k = \frac{sl}{K}$$

where s = slip
l = length of ground contact of traction device (wheel or track)
K = soil deformation modulus

(Reproduced by permission of Elsevier Scientific Publishing Co.)

and each wheel has a contact area of $0.11 \, m^2$ and contact length 400 mm. It operates on a soil with characteristics $c = 5 \, kN/m^2$, $\phi = 29°$, $K = 35 \, mm$. Find the slip when $H = 8 \, kN$. Ideal maximum thrust from Micklethwaite equation

$$H_{max} = Sc + Q \tan \phi$$
$$= 2 \times 0.11 \, m^2 \times 5 \, kN/m^2 + 26 \, kN \times \tan 29°$$
$$= 1.1 \, kN + 14.4 \, kN = 15.5 \, kN$$
$$a = H/H_{max} = 8 \, kN/15.5 \, kN = 0.516$$

from graph Figure 6.20 corresponding value of $J_k = 1.6$

$$J_k = 1.6 = \frac{sl}{K} = \frac{s400 \, mm}{35 \, mm}$$

$$s = \frac{35 \times 1.6}{400} = 14\%$$

The traction theory outlined is an approximation to a complex interaction involving both tyre and soil characteristics. It cannot be relied upon to give an accurate prediction of performance in a particular case. Nevertheless as a general model it provides a good understanding of the effects on tractive performance of particular factors, for example contact length. It will be seen from the example above that the slip corresponding to a given thrust ratio varies inversely with the contact length. A 50 per cent increase in contact length, from 400 mm to 600 mm, would result in a reduction in slip to $\frac{2}{3} \times 14\% = 9.3\%$. In practice an increase in contact length may be obtained by fitting tyres of larger diameter and/or by operating them at the lowest permissible pressure to increase the contact area, and hence contact length.

The effect of the contact length factor may be pursued further in order to investigate the slip at which maximum power is developed at the driving wheels. At zero slip no soil deformation is taking place, hence no thrust is being produced and in consequence no power. At 100 per cent slip considerable thrust may be generated but, since there is no forward movement of the tyre, again no power is being produced. Between these extremes, assuming a constant input speed to the driving wheels, the power generated at the wheels rises to a maximum. The slip at which the maximum power is produced depends on the ratio $\frac{l}{k}$ and Figure 6.21 shows the manner of this variation. The importance of the length of the contact patch is emphasized by this relationship and since the length is a function of wheel diameter ($l \approx 0.31 \, d_c$) it will be seen that increasing the wheel diameter will result in improved slip performance. From the point of view of tractive performance the maximum feasible wheel diameter should be used whenever alternatives are available, or permissible.

The maximum drawbar power is produced by a wheeled tractor at a slip of about 20–25 per cent, and by a tracklaying tractor at about 10–12 per cent. In practice a slip of 20 per cent would normally be regarded as too high because of potential harmful effects to soil structure—12–15 per cent is a more reasonable

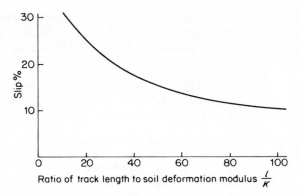

FIGURE 6.21 Slip at which maximum power is produced
at the wheel or track as a function of l/k

maximum in working conditions. The thrust H which a wheel or track produces is related to the ideal maximum thrust H_{max} (see Figure 6.20) whilst H_{max} is itself a function of the load Q on the wheel or track, the contact area S, and the soil strength characteristic c and ϕ (Eqn. 6.12). The load Q varies with tractor operating conditions, in particular it is increased by weight transfer and weight addition effects when mounted implements are in use. Also, of course, it has the same magnitude as the vertical soil reaction Z at the drive wheels or tracks. Taking these factors into account it is useful to define a thrust coefficient, so that

$$\mu = \frac{H}{Q} = \frac{H}{Z} \qquad 6.13$$

6.6.4 Rolling resistance

The pull R required to move a wheel or track across a horizontal surface is called the rolling resistance. Rolling resistance tends to increase in direct proportion to an increase in the load Q, which the wheel is supporting. Therefore the ratio $\dfrac{R}{Q}$ is a constant for a particular wheel or track operating over a particular surface and is called the coefficient of rolling resistance ψ, so that:

$$\psi = \frac{R}{Q} = \frac{R}{Z} \qquad 6.14$$

The rolling resistance R may be calculated from the equation

$$R = \frac{0.15\,b}{l}\,Q \qquad 6.15$$

The rolling resistance of a wheel with a pneumatic tyre has two main components: the internal rolling resistance which is caused by loss of energy resulting from the continuous flexing of the tyre carcass as the wheel rotates in contact with the ground; and the external rolling resistance arising from the energy which the wheel has to expend in deforming the soil surface. In off-highway conditions the rolling resistance caused by soil deformation is much greater (five times or more) than the internal resistance due to tyre or track construction.

A rut formed by the passage of the wheel is evidence of the expenditure of energy in deforming the soil and in general the larger the cross-section of the rut the greater the rolling resistance. Sinkage in a particular soil may be reduced by reducing the ground contact pressure of the wheel; this may be achieved for a given load by increasing the size of the contact area either by using larger size tyres or decreasing the inflation pressure to the permissible minimum. It should be borne in mind that it is preferable to increase the length of the contact area by choosing a larger diameter tyre rather than to increase the section width of the tyre. For a given contact area the depth of sinkage would be similar whether the area was wider or longer. However, it should be remembered that the width of the rut would be greater with the wider contact area, leading to higher rolling resistance.

It should be noted that when the inflation pressure of a tyre is reduced not only is the contact area increased but so also is the deformability of the tyre relative to the soil (Figure 6.22). This results in more uniform depth of sinkage along the length of the contact area and consequently less depth of rut and a shallower rolling resistance. Much of the quoted data on the rolling resistance of wheels derives from work performed by McKibben and Davidson in 1939 and this information is presented in Figures 6.23 and 6.24 in a readily usable format. It is important to reduce the rolling resistance to a minimum since:

$$F_x = H - R$$

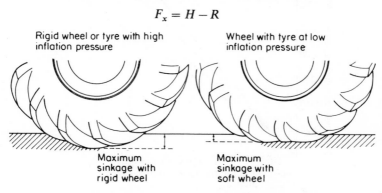

FIGURE 6.22 Sinkage pattern of rigid and deformable wheels

meaning that the rolling resistance detracts from the thrust produced at the driving wheels, resulting in less capability for the production of drawbar pull.

6.6.5 Steering

To provide steering effect the driver sets the steered wheels at an angle δ to their direction of motion. A lateral force Y is generated which acts on the wheel from

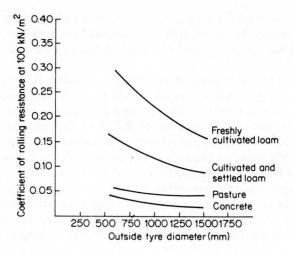

FIGURE 6.23 Effect of wheel diameter on coefficient of rolling resistance

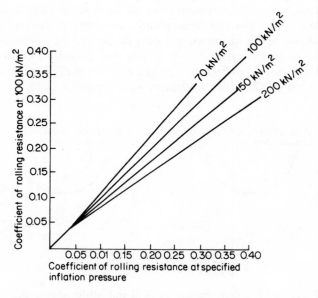

FIGURE 6.24 Effect of inflation pressure on coefficient of rolling resistance

the soil. Since the soil must be deformed in order to provide this force the wheel will have a sideways component of motion as well as the motion in the direction of the plane of the wheel. Thus its actual direction of motion will be v_a and not the direction indicated by the angle of turn v_t. The difference between these two directions is the angle of drift ε of the wheel. In automobile engineering terminology this angle is called the slip angle but the term drift angle is preferred for off-highway vehicles to avoid confusion with wheelslip due to thrust which, if the wheel is being driven, may be occurring simultaneously with drift. As with slip/thrust and rolling resistance the predominating influence on the drift angle is deformation of the soil; deformation of the tyre is usually of minor consequence: see Figure 6.25.

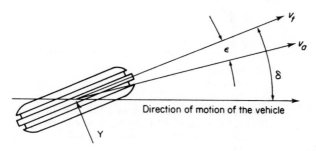

FIGURE 6.25 Motion of a steered wheel

The analysis used to determine the slip/thrust relationship may be extended to deal with both slip and drift for driven and steered wheels. This analysis shows that the angle of drift is a function of the ratio:

$$\mu_y = \frac{Y}{Q} = \frac{Y}{Z}$$

The relationship is illustrated for a particular case by Figure 6.26. When the wheel is driven the effect is to increase the angle of drift, for a given side force, as the thrust from the wheel, and hence the slip, is increased (Figure 6.27.) For steerability the drift angle ε should be only a fraction of the steering angle δ.

When $\varepsilon = \delta$ the wheel will continue to move straight ahead despite its being steered away from the straight-ahead position. Reasonable control is possible

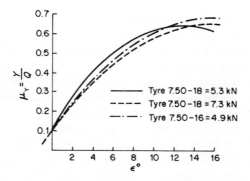

FIGURE 6.26 The ratio μ_y as a function of ε for two wheels

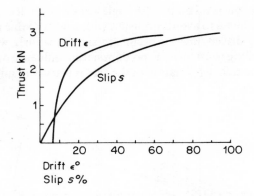

FIGURE 6.27 Slip and drift as a function of
thrust for a wheel with constant side force
(μ_y = constant) on a particular soil

when ε is not more than about $\frac{1}{3}\delta$ but different operators display varying degrees
of tolerance to difficult steering situations. The analysis shows that steerability
depends on an adequate load on the steered wheels. If a four-wheel-drive tractor
is steered while pulling its steerability is likely to be diminished if the steered
wheels are also contributing a substantial pull. A two-wheel-drive tractor
working on a side slope will also experience difficulties if the wheelslip is excessive,
since the rear wheels will tend to drift downhill.

Most undriven steered wheels in off-highway use have pronounced circumferential ribs to assist in providing sidethrust—an effect which is not taken
account of in the discussion above but which depends on a sufficient loading on
the wheel to force the ribs into the ground. The increased contact pressure under
the rib also helps in penetrating a slippery surface layer.

6.7 SOIL COMPACTION AND SMEAR

One main objective of soil cultivation is to loosen soil, allowing good aeration
and water status for optimum plant growth. Loosened soil is subsequently
compacted by the natural processes of weathering but additional compaction,
harmful to growth, may result from the movement of tractors and machinery over
the surface. Excessive compaction may lead to: poor aeration; impeded drainage;
resistance to root penetration; and clod formation causing harvesting and
separation difficulties, particularly with potatoes.

A mass of soil consists of solid particles separated by voids which may be
partially or completely filled with water, the remainder of the spaces being taken
up by air. The size of the solid particles varies from clay particles with a nominal
diameter of 0.002 mm through silt (0.002 mm to 0.06 mm) to sand (0.06 mm to
2.0 mm) with larger particles classified as gravel. The proportion of each of the
particle sizes determines the texture of the soil which may be described as a loamy

sand, clay, silt loam, etc., and affects among other things the ease of cultivation and the growth of the plant. The porosity n of the soil is defined as the volume of voids V_v compared to the total volume of the soil V

$$n = \frac{V_v}{V} \hspace{3cm} 6.16$$

Porosity is significant in its influence on good plant growth through root development. A porosity of around 50 per cent with about half of the pore space taken up by water, giving an air space of about 25 per cent, provides a good basis for plant development; although optimum conditions may vary considerably for particular crops and soils. As a general rule plant growth and yield are likely to suffer appreciably if the air space is reduced below 10–15 per cent depending on the crop grown.

Movement of a wheel or track over the ground surface creates a pattern of stress within the soil mass, caused by the compressive and shearing stresses at the contact area and dependent on various characteristics of the soil. Stresses at the contact area result from the load which the wheel is supporting and the tractive and steering action which the wheel is providing. The distribution of compressive stress in the soil under a wheel has been investigated by a number of research workers both theoretically and experimentally. Figure 6.28 compares the stress distribution under a single load wheel and dual wheels carrying the same load.

The compressive stresses arising in the soil cause compaction, that is a reduction in volume which is entirely at the expense of the volume of voids. The reduction of volume, with an associated increase in density, continues until the solid particles are forced into a close-packed state and cannot be compacted further by compression alone, although vibration or shearing action may cause further reduction in volume.

Figure 6.29 shows how air is first squeezed out, thus reducing the volume. Water will then be squeezed out if its volume is greater than the close-packed voids volume shown at A. If the water volume is less than this it will remain in the soil, together with any air necessary to take up the remainder of the voids space. Maximum compaction is likely to occur when the volume of water is slightly less than the void volume at A, since the water will lubricate the solid particles, assisting their close packing. If the water volume is greater than at A (as shown in the figure) the passage of the wheel may be so rapid that the water does not have time to be squeezed out but instead will provide buoyancy to help carry the load. In the worst conditions for compaction, mentioned above, just one pass of a heavily loaded tractor wheel may cause up to 90 per cent of the maximum compaction. The full maximum may be achieved with only a few more passes.

Since soils are known to show markedly different responses dependent on the duration and strain rate associated with the given applied stress, compaction under wheels is likely to be related to vehicle speed. Figure 6.30 shows the effect of forward speed for a particular wheel which shows that the compaction effect is less if the tractor is travelling fast.

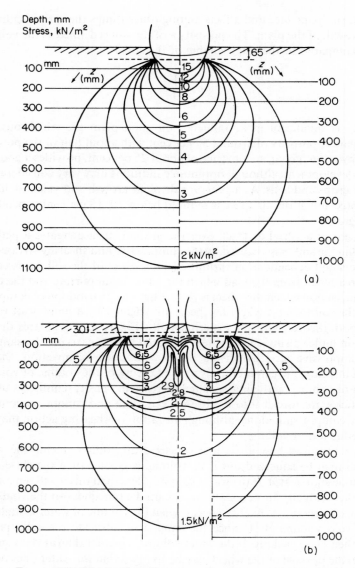

FIGURE 6.28 Distribution of compressive stresses under (a) single wheel and (b) dual wheel carrying the same load

Compaction damage may be minimized by: crop production techniques involving the least passage of tractors and machinery; working when soil moisture conditions are favourable; avoiding high ground contact pressures. Figure 6.31 shows some of the options available for reducing compaction in relation to the factors affecting the cultivation system as a whole.

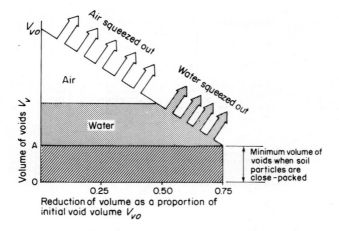

FIGURE 6.29 Reduction of volume as a proportion of initial void volume, V_{vo}

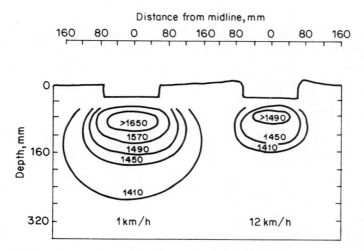

FIGURE 6.30 The effect of forward speed on the distribution of soil bulk density (kg/m³) below a tractor wheel (after Soane, Dickson and Campbell, 1982. Reproduced by permission of Elsevier Scientific Publishing Co.)

Ground contact pressures may be kept to a minimum by reducing the weight on the wheels and increasing the contact area. Weight may be reduced by removing ballast to the minimum compatible with low wheelslip. Contact area may be increased by deflating the tyres to the minimum permissible pressure for the load carried; or by increasing tyre size to give greater diameter and/or width; or by fitting dual wheels. An inflation pressure of 80 kN/m^2 is the maximum

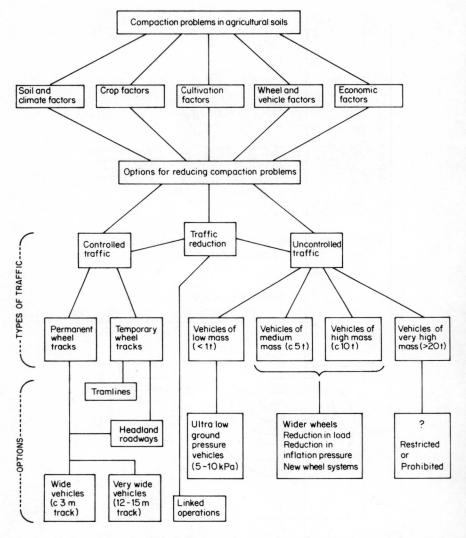

FIGURE 6.31 A simplified diagrammatic representation of some of the options available for reducing compaction in relation to the factors affecting the cultivation system as a whole (after Soane, Dickson and Campbell, 1982. Reproduced by permission of Elsevier Scientific Publishing Co.)

advisable if soil compaction is likely to be a problem. Figure 6.28 illustrates the very significant effect of dual wheels in reducing compressive stresses and the penetration of these stresses into the ground. Their use is strongly recommended when soil compaction to an undesirable level is likely.

Compaction in itself is serious enough but the situation is made even more harmful by the effects of smear. When a heavily loaded traction wheel is operated in high moisture content soils it will work at a high slip which kneads the soil particles together to form a homogeneous mass—like clay worked on a potter's wheel. When it dries out later in the season the result is a layer of soil which is almost impenetrable to roots, air, and water, causing a reduction in crop yield. Again, the practical solution is to operate with the largest possible tyres or dual wheels at a low pressure and a light load both vertically and tangentially. At the limit it would be better not to go on the land at all when the moisture content is critical, particularly in soils with a high clay content.

6.8 THE EFFECT ON TRACTIVE PERFORMANCE OF TRACTION AIDS

In many field operations the travel speed of a tractor is limited since, due to problems of implement design and control, the effectiveness of the operation deteriorates with increasing speed. A limitation also arises from the effects of noise and vibration, which may become intolerable to the operator at higher speeds. Consequently most field operations are undertaken at a maximum speed between 2 m/s and 3 m/s (7 km/h and 11 km/h or 4.5 mile/h and 7 mile/h).

In order to use the full power of a tractor at this speed high drawbar pulls are required. For example a 50 kW (67 hp) tractor working at 3 m/s must ideally produce a pull of 17 kN (see Eqn. 6.1) in order to utilize all its available power, whilst at 2 m/s a pull of 25 kN would be necessary. Since the total weight of such a tractor is about 28 kN it would be able to generate such pulls only with excessive slip, if at all, and consequently with very poor efficiency.

Various traction aids are available to improve the potential pull of a tractor and hence reduce the slip for a given pull. It should be borne in mind that in increasing drawbar pull many traction aids also increase rolling resistance. As a result overall efficiency of the wheel may be reduced with consequent increase of fuel consumption for work done.

TABLE 6.2 Characteristics of Some Traction Aids

	Increased contact area	Increased contact length	Increased rolling resistance	Increased compaction	Penetration to stronger soil	Suitability for use on roads
Diff lock						0
4-wheel drive	x	x				xx
Half-tracks	xxx	xx			x	0
Dual wheels	xx					x
Ballast			xx	xxx		xxx
Strakes		xxx			xxx	0
Cage wheels	xx		xx			x
Chains			x		x	0

0	not recommended
x	low
xx	moderate
xxx	high

TABLE 6.3 Selection of Traction Aids for Various Operations and Soil Conditions

		Type of Operation		
		Heavy draught		Medium draught	
	In furrow	On the land	Loose soil	Settled soil
Typical operations	Mouldboard ploughing (low and medium power tractors)	Mouldboard ploughing (high power tractors) Chisel ploughing (all tractors) Subsoiling	Seedbed preparation Sowing Fertilizer didtribution	Fertilizer distribution
Clay soil (cohesive)	1. 4-wheel drive 2. Wheelstrakes 3. Diff-lock	1. 4-wheel drive 2. Half track 3. Dual wheels 4. Wheelstrakes	1. Half track 2. 4-wheel drive 3. Dual wheels 4. Cage wheels	1. 4-wheel drive 2. Dual wheels 3. Wheelstrakes
Sandy soil (frictional)	1. Ballast 2. Diff-lock	1. Ballast (and dual wheels) 2. 4-wheel drive	1. Half-track 2. 4-wheel drive 3. Dual wheels (and ballast) 4. Cage wheels	1. 4-wheel drive 2. Dual wheels (and ballast)
Loam soil (cohesive-frictional)	1. Ballast 2. Wheelstrakes 3. Diff-lock	1. Ballast (and dual wheels) 2. 4-wheel drive	1. Half-track 2. 4-wheel drive 3. Dual wheels (and ballast) 4. Cage wheels	1. 4-wheel drive 2. Dual wheels (and ballast) 3. Wheelstrakes
Soil with slippery top layer	1. Wheelstrakes 2. Diff-lock 3. Chains	1. Wheelstrakes 2. Chains 3. Half-tracks	1. Half-tracks 2. 4-wheel drive	1. Wheelstrakes 2. Chains 3. Half-tracks

Note: Where 'Ballast (and dual wheels)' is suggested dual wheels should be fitted:
[a] if soil compaction is likely to occur
[b] if ballast required exceeds the capacity of single wheels fitted.
Where 'Dual wheels (and ballast)' is suggested the dual wheels alone should usually provide the additional traction required but ballast may be added if necessary.

Table 6.2 summarizes the main characteristics of some of the more common traction aids and Table 6.3 shows the recommended aid for a particular situation.

6.8.1 Traction aids

Traction may be improved by the use of suitable traction aids and, in most circumstances, by ballasting to increase the load on the driven wheels. Ballasting normally means adding cast iron weights, either on the chassis and/or on the wheels, but in circumstances where permanent weight addition is acceptable the weight of a wheel can be increased considerably by filling or partially filling its tyre with a liquid, usually water as it is cheap and readily available. One litre of water weighs 9.8 N (1 kgf) so that, for example, a 13.6–36 tyre will provide an additional ground contact load of about 2 kN when 75 per cent filled with water. 75 per cent filling is achieved when the level of water is up to the valve at its highest (12 o'clock) position. Higher degrees of filling are possible, but special valve and pumping equipment is then necessary, the tyre casing becomes more vulnerable to impact damage, and ride characteristics may be affected. When calculating tyre loading the additional weight of the liquid must be included and the maximum carrying capacity of the tyre must not be exceeded. If this happens a tyre with a higher rated load must be used to obtain the required performance, that is of increased ply rating or increased size, or alternatively, dual wheels may be considered. In climates where frost is likely it is necessary to guard against freezing. Most manufacturers recommend flake calcium chloride (70 % CaCl$_2$) as the best cheap material to provide an anti-freeze solution. Common salt (NaCl) is too corrosive unless used with a corrosion inhibitor and is not able to provide protection below −18°C (0°F). Ethylene glycol, used in commercial anti-freeze for water-cooled tractor engines, provides excellent protection but is expensive relative to calcium chloride. Table 6.4 shows the solution strengths required to provide alternative levels of frost protection. Because the volume of air is reduced by liquid filling the inflation pressure must be checked more frequently. A special gauge for water-filled tyres must be used and the valve must be at its lowest position when the pressure is taken so that the pressure recorded includes the pressure due to the head of water in the tyre. Full details of the procedure for filling tyres with liquid and tables of quantities of water and anti-freeze agent required for various tyre sizes may be obtained from tyre distributors or from 'Care and maintenance of agricultural tyres', published by the Service Department, Dunlop Ltd., Fort Dunlop, Birmingham.

TABLE 6.4 Anti-freeze Recommendations for Water-ballasted Tractor Tyres

Anti-freeze agent	Solution strength required for protection	
	Down to −7°C (20°F)	Down to −18°C (0°F)
Flake calcium chloride (70 % CaCl$_2$)	130 g to each litre of water (1.30 lb/gal)	240 g to each litre of water (2.40 lb/gal)
Ethylene glycol	1 volume of ethylene glycol to 4 volumes of water	1 volume of ethylene glycol to 2 volumes of water
Common salt (NaCl)[a]	125 g to each litre of water (1.25 lb/gal)	340 g to each litre of water (3.40 lb/gal)[b]

[a] A recommended inhibitor must also be used.
[b] At this concentration the solution is fully saturated and it is difficult to dissolve this quantity of salt.

6.9 DYNAMIC CHARACTERISTICS OF TYRES

Previous discussion has related to the characteristics of tyres under steady states of loading, and Figure 6.9 showed the static load deflection curves as a lattice plot. The dynamic characteristics are slightly different, the dynamic stiffness being on average 10 per cent greater than static for rear traction tyres. At high pulls and low speed, when the tangential load due to traction is of the same order as the vertical load, the dynamic stiffness is reduced by about 10 per cent. These characteristics are not significantly changed by water ballasting. Old tyres have a lower stiffness, often 25 per cent less.

The tractor, supported on its tyres, acts as a simple vibrating system. If the tractor is made to vibrate it will do so at a frequency determined by its mass and the stiffness of the tyres. This frequency is called the natural frequency. The natural frequency of rear traction tyres is in the range 2.2–4.3 Hz and of the smaller front tyres 3.3–4.5 Hz. If the tractor hits a bump it will start to vibrate at its natural frequency. The vibrations die away at a rate determined by damping capacity of the tyres. With high damping the natural vibration is soon reduced—this is called a high damping ratio. With a low damping ratio, the vibrations take a long time to die away (Figure 6.32).

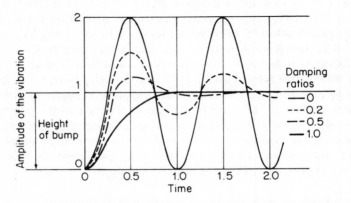

FIGURE 6.32 Amplitude-time curves showing the effect of damping ratio

Tractor tyres have a damping ratio of 0.06–0.08 with a maximum of 0.1. Much higher damping ratios of about 0.7 are required (a typical value for a car suspension) to damp out the vibrations quickly. However, tyre construction prevents changing the damping ratio or stiffness up to the required value. The surface that a tractor travels on is usually very bumpy. Each successive bump causes the tractor to vibrate at its natural frequency. If frequency of bumps on the surface is the same as the natural frequency of the tractor the situation is aggravated, and amplitude of the vibration builds up, resulting in excessive bouncing which makes tractor control difficult. The bouncing, particularly on high-speed road work, can cause damage to the tyre carcass from excessive deflections and from heat build up, causing the parts of the tyre to separate.

The driver also suffers from vibration effects. Figure 6.33 is taken from an international standard which shows that, in the frequency range 4–8 Hz, his body can tolerate only low levels of vibration, this being the range of natural frequency

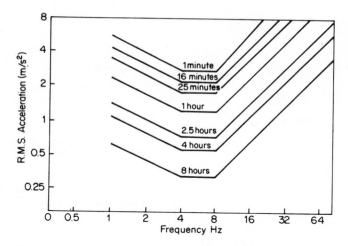

FIGURE 6.33 Fatigue-induced proficiency boundary for vertical vibration as a function of frequency and response time

of his abdominal cavity and organs. Long exposures at this frequency range can cause physical harm to the driver.

The solution presently adopted to improve the driver's comfort is to provide a sprung and damped seat. Due to the magnitude of the mass involved (of the driver) the spring rate must be very low, resulting in large deflections which cause control problems and also introduce roll (sideways) movements with the tendency to transfer the discomfort to another direction. Tractors fitted with a complete suspension system are not yet available, due to the cost of adapting to a wide range of load and implement control requirements.

6.10 STEERING

Mention has already been made of the forces that can be developed by the steered wheels on the soil which can be used to make the vehicle travel in the required direction.

Small single-axle tractors are steered by the operator simply pushing sideways on the handles. The larger types of these tractors require steering clutches and brakes as the machine is too heavy and powerful to be steered by the operator alone. The technique is usually to disengage the drive to the wheel on the inside of the curve and apply the brake on that side.

A single-axle tractor designed at IITA uses a free wheel on one side and a V belt and small pully reverse system operated by a 'deadman's' lever on the handlebar to drive the fixed wheel backwards round the free wheel. This system is very easy to operate and requires very little effort: see Figure 6.34. When the tractor is made into a powered trailer (see Chapter 7) the steering pivot should be arranged to be as close to the tractor axle as possible and the tractor wheels set as closely as

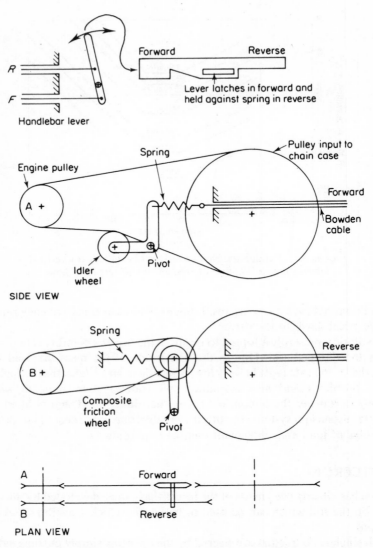

FIGURE 6.34 Single-axle tractor V-belt forward and reverse system

possible to reduce the sideways load on the handlebars when going over rough ground. Substantial steering stops are essential to prevent the machine from steering too sharply, which will either cause instability or cause the wheels to touch the drawbar, or both.

Most conventional tractors have four wheels with the two front wheels steering. There are two aspects of steering control that should be considered; one is the problem of making the tractor travel more or less straight when pulling a load, and the other is making it steer to go in the required direction.

Drawbar load has an effect on the steering. In order to obtain maximum traction performance it is necessary to maximize the weight on the drive wheels (see Chapter 7). In doing so the weight on the steer wheels will be small so their side force capability will be low. This in turn means that the resultant drawbar force from the implement must be more or less along the centre line of the tractor, otherwise the turning moment of the force will be too great for the wheels to resist. Figure 6.35 shows the effect of the offset drawbar load on the side force on the front wheels.

$$\text{For equilibrium } Y = \frac{Fe}{w} \qquad\qquad 6.17$$

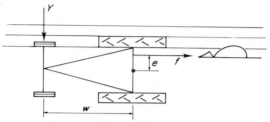

For equilibrium $Y = \frac{Fe}{w}$, The greater the offset e the greater the side force Y.

FIGURE 6.35 Effect of an offset drawbar load, F, on the front wheel side force, Y

If $\frac{Fe}{w}$ is greater than μZ then the front wheels will slide sideways. Even lower forces, however, will cause the front wheels to drift sideways. The driver must steer to the left (in this case) to enable the tractor to continue in a straight line. The result is a larger rolling resistance, excessive tyre wear, and fatigue of the operator. To overcome the problem the effective drawbar load must be along the centre line of the tractor: in this case it is necessary to reduce the wheel spacing t by a certain amount. (All tractors have wheel width track adjustment systems for this reason.) The requirement to fit the various row crop widths is secondary as it is important to select a tractor that can be loaded symmetrically when straddling a particular number of rows.

The other aspect of tractor steering is to enable it to go in the required direction. Most high-draught operations are carried out in more or less straight lines. This is partly for convenience in the field but also because the angled drawbar force which results from travelling in a curve has too great an influence on the steering performance of the wheels.

Figure 6.36 shows the effect of the drawbar force on the steering

$$Fe' = WY \cos \delta$$

Since F is usually large, e' must be small for reasonable steering control. For e' to

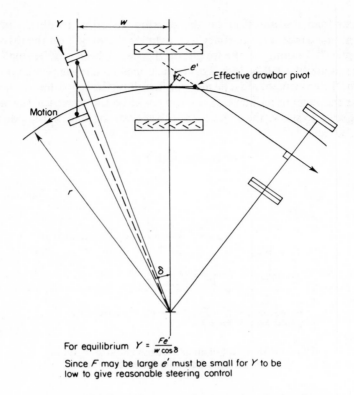

For equilibrium $Y = \dfrac{Fe'}{w \cos \delta}$

Since F may be large e' must be small for Y to be low to give reasonable steering control

FIGURE 6.36 Effect of drawbar force on the steering force

be small the effective drawbar pivot must be close to the back axle centre line. On a small radius of turn there is a danger of the drawbar fouling the tyres.

A three-point-mounted implement has an effective drawbar pivot in front of the back axle and this will actually assist the steering effect, making control very light and easy. For a small turning radius however, the check chains limit the sideways movement (to prevent the linkage touching the tyres) and the tractor implement combination cannot in general be steered at all after this point is reached. The solution to this small radius steering problem is to lift or get rid of the drawbar load by some means when turning sharply.

6.10.1 Steering geometry

It is necessary for certain design features to be incorporated into the design to give reasonably accurate steering.

Figure 6.37 shows the castor angle and the king pin inclination angle. The principle is that the centre line of the steering pivots or king pins when produced should touch the ground near the front of the contact patch and near its middle. This will give the required steering characteristic to the vehicle, that is it will tend

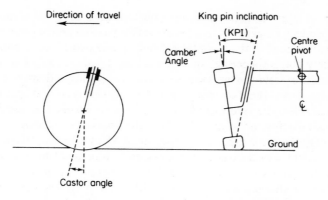

FIGURE 6.37 Castor and king pin inclination angles of steered
wheels

to travel straight and rough ground will not cause too much kick-back or shock
loads into either the steering linkage or the operator through the steering wheel.
The angles are usually only a few degrees, the castor angle being reduced or even
zero for front-wheel-drive machines.

With the normal linkage—basically an Ackerman system with Jeantaud
modification—there is only a limited range of steer angle when the wheels have an
approximately common steer centre.

Figure 6.38 shows the commonly used steering linkage when the track rod is

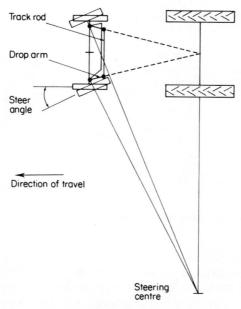

FIGURE 6.38 Steering linkage

slightly shorter than the distance between the king pins. This makes the line of the drop arms produced meet at $\frac{2}{3}w$ to w from the front axle, where w is the wheel base. The wheels also toe in at the front by 2–4 mm. These two effects combine to give approximately correct steering for a tractor up to about 30°. Except for one particular angle of steer the tyres are forced to drift slightly sideways, particularly on sharp turns, but as the tractor is going relatively slowly on soft ground (or on a hard dusty surface) the tyre wear is usually kept to within reasonable bounds.

If the wheel spacings are changed it is usually necessary to change the width of the track rod at the same time. The earlier small Massey Ferguson tractors had a radius rod system which avoided the track rod adjustment mentioned.

Small tractors may use a steering tiller for the operator. This is very simple and works well as the steering forces are small. Medium-sized tractors use a steering box to gear down the movement on the steering wheel to give about $3\frac{1}{2}$ to $4\frac{1}{2}$ turns lock to lock. This will keep the operator's steering effort down to acceptable levels.

Large tractors are fitted with power steering, usually of the fully hydrostatic type powered by an engine-driven pump. The most common arrangement is using cross-linked, double-acting cylinders as shown in Figure 6.39. If the engine fails it is theoretically possible to steer the tractor but in practice it is very difficult. The pump steering wheel unit and cylinders are selected to give the required steering forces at about 3 to 4 turns lock to lock.

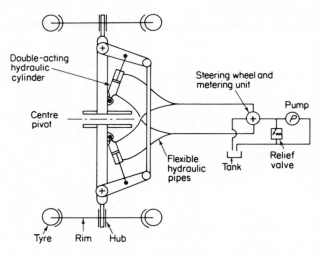

FIGURE 6.39 Cross-linked, double-acting cylinder steering system

6.11 BRAKES

Most tractors have brakes on the rear wheels only and these can be operated by foot pedals together for use in the transport mode or separately for use as steering brakes to assist the front wheels in turning sharp corners. A parking brake is also connected to the brakes through the same or a separate linkage. At the present time a retardation of 0.25 g with a 600 N pedal load is considered an adequate performance although many manufacturers manage this retardation at much lower, more reasonable, pedal loads. (g is acceleration due to gravity).

Small tractors often have simple band brakes which are relatively cheap to make and easy to operate (see Figure 6.40). However, larger tractors use disc brakes, having clutch-like discs as distinct from car-type disc brakes. These tractor brakes are totally enclosed and are operated by a ball and cam ring system giving a good servo action which results in high brake performance in a small space with low operating forces (see Figure 6.41). Drum brakes and car-type disc

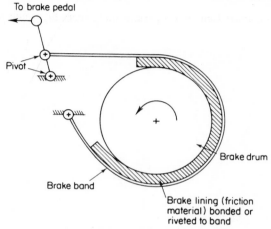

FIGURE 6.40 A simple band brake

With the rotation and brake pedal forces in the directions shown the brake will have a strong servo action—the rotation helps to apply the brake thus giving a greater effect for a given pedal effort

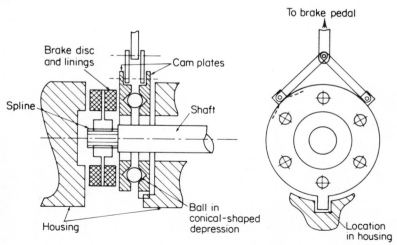

FIGURE 6.41 A tractor disc brake

brakes are not used very widely on modern tractors because of difficulties of sealing against adverse conditions.

REFERENCES

Soane, B. D., Dickson, J. W., and Campbell, D. J. (1981.) Compaction by Agricultural Vehicles: A Review. *Soil and Tillage Research*, **1** (1980/1981).

ACKNOWLEDGEMENT

Thanks are due to Dunlop Limited for permission to reproduce material in this chapter.

Appendix: List of symbols

F	Force at a tractor drawbar, arising from implement pull
F_x, F_y, F_z	Longitudinal, transverse and vertical components of a drawbar force F, see Figure 5.1
G_t	Weight of a tractor
H	Gross thrust force acting on a tractor driven wheel
H_{max}	Maximum gross thrust force
J_k	Slip ratio (deformation ratio) $= \dfrac{sl}{k}$
K	Soil deformation modulus
P_d	Power transmitted to the ground drive (wheels or track) of a tractor
P_e	Power developed by a tractor engine, measured at the fly-wheel
P_F	Drawbar power
Q	Load on a wheel
R	Rolling resistance of a wheel
S	Contact area of a wheel on the ground
V	Total volume of a soil sample
V_v	Volume of the void spaces in soil sample
Y	Lateral force on a wheel
Z	Normal component of the soil reaction at a wheel
Z_f	Normal soil reaction at a front wheel
Z_r	Normal soil reaction at a rear wheel
b_c	Catalogue width of a tyre
c	Soil cohesion
d	Actual travel distance of a wheel
d_c	Catalogue diameter of a tyre
d_n	'No slip' travel distance of a wheel
d_r	Diameter of a wheel rim

e	Drawbar offset from centre-line
h	Section height of a tyre
j	Soil displacement
j_{max}	Maximum displacement of the soil in contact with a wheel or track
l	Length of the ground contact area S ($\approx 0.31\, d_c$ for a wheel)
n	Soil porosity $= V/V_v$
r_s	Static loaded radius of a wheel
s	Slip of wheel or track
t	Tractor width
v	Actual speed of travel
v_n	'No slip' speed of travel
w	Tractor wheel base
α (alpha)	Thrust ratio $= H/H_{max}$
β (beta)	Angle of slope of a ground surface
δ (delta)	Turning angle of a steered wheel
ε (epsilon)	Angle of drift of a wheel
η_F (eta)	Gross tractive efficiency of a tractor $= P_F/P_e$
η_R	Force ratio F_x/H defining rolling resistance efficiency
η_s	Velocity ratio defining slip efficiency $(v_n - v)/v_n$
η_t	Power ratio P_d/P_e defining transmission efficiency
μ (mu)	Thrust coefficient of a wheel $= H/Z$
μ_Y	Lateral force coefficient of a steered wheel $= Y/Z$
σ (sigma)	Normal stress on a soil surface
τ (tau)	Shearing stress on a soil surface
τ_{max}	Maximum or limiting stress on a soil surface
ϕ (phi)	Angle of internal soil friction of a soil
ψ (psi)	Coefficient of rolling resistance of a wheel R/Q

7 *Characteristics of Implements*

A reversible disc
plough operating in
hard dry soil in
Central Africa

In order to perform an agricultural operation it is necessary to select an implement of the right size to do the required task, and to attach it to the power unit in such a way that it can be controlled and will be within the power capacity available. The aim of this chapter is to describe the main characteristics of a range of possible implements which may be available for small-scale mechanization.

The various agricultural operations can be classified as follows:

(1) *Primary cultivation.* The working of undisturbed soil to loosen it to the required depth and/or to bury trash and to control weeds. Making the major soil erosion earthworks and land levelling may also be considered under this heading.

(2) *Secondary cultivation.* The working of the previously loosened soil into the required clod size and distribution and providing the correct degree of compaction to give good soil contact with the seed or plant with the correct permeability to air and water.

(3) *Planting and transplanting.* The placing of the seed and seedling at the correct depth and spacing in the soil.

(4) *Crop upkeep operation.*

(a) Spraying—the chemical control of weeds and other pests such as insects, fungus diseases, etc.

(b) Fertilizing—the provision of the correct balance of plant nutrients in the soil.

(c) Weeding—the controlling of weeds to avoid significant competition with the crop.

(5) *Harvesting.* The collection of the required part of the crop for subsequent use.

(6) *Processing.* The work done on the raw crop to convert it into a storable, usable or saleable item.

(7) *Transport.* The moving of material to the required place at various stages in the process:

e.g. production input materials (seeds, fertilizers and insecticides) and machines to field;

crop to farmstead;

product to market.

There are many types and sizes of farm implements available which have been produced to provide ways of doing these tasks more easily, quickly and cheaply than can be done by human hand alone. Many of the designs have been developed over considerable periods of time to suit particular human, soil, and crop conditions. It is not possible to describe here all the details and conditions but each major type of implement has certain basic characteristics and adjustments which affect its performance.

7.1 TRACTOR PLOUGHS

The four major types of implements used for primary cultivation are mouldboard, disc, chisel, and rotary ploughs.

7.1.1 Mouldboard plough

The mouldboard plough is designed to cut, lift, and invert the soil, to bury trash and weeds and leave a bare soil surface. Its cutting unit, called the bottom, consists of the share, mouldboard, and landside: the share is provided with the lead to land and the pitch as shown in Figure 7.1. The landside stabilizes the bottom by counteracting the side thrust exerted by the soil on the mouldboard. In some tractor mouldboard ploughs the landside is replaced by a bar furrow wheel set at an angle with respect to the direction of travel. A knife or disc coulter may be installed ahead of the bottom to make the vertical cut in the soil to enable the furrow slice to be separated and easily turned from the unploughed land.

The width of cut of a bottom is the perpendicular distance from the landside to the next landside on the adjacent body. Single-furrow ploughs must be controlled

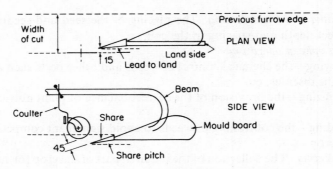

FIGURE 7.1 Basic parts of a mouldboard plough

to run at this corresponding distance by adjusting the linkage: on simple animal-powered ploughs this is done by eye.

Smaller-sized ploughs often have a simple knife coulter or no coulter at all, as for example on ox ploughs, for soft conditions. Larger, heavy-duty ploughs for tractors may be fitted with bar points and may have various impact-protection devices such as shear pins, spring-loaded beams, or hydraulic accumulators.

7.1.2 Disc plough

The disc plough utilizes a concave disc which rotates during operation and is set at an angle in relation to the direction of travel and to the ground surface. Disc ploughs are more suitable for hard soil containing obstructions but they are not usually made in the smaller sizes as they tend to be expensive and more difficult to maintain. Polydisc ploughs have more and smaller discs for wide shallow work at high speeds behind relatively large tractors. The large side thrust from the soil is carried on an angled steel wheel, usually of the disc and flange type, at the rear.

For a trailed plough it is necessary to adjust the drawbar in such a way that the plough will work at the required depth and pull straight. For instance, if the plough front depth wheels sink in badly and the plough tends to pull towards the ploughed land, the plough end of the drawbar should be moved downwards, and the tractor end across towards the unploughed land.

With a mounted plough the cross-shaft, with the crank down on the ploughed side, can be angled by a screw adjustor. This will change the front furrow width. If the tractor steering is affected too much the shaft can be slid slightly sideways to reduce the effect. The front furrow on reversible ploughs is usually adjusted by stops on the right-hand end of the cross-shaft, the front stop being for one way and the rear stop for the other; they should be adjusted so that the furrows match both ways. For ploughs particularly, and also some other implements, it is important to set the tractor wheels at the correct width for the plough size, otherwise the adjustments referred to will be ineffective and poor work will result.

Ridging bodies can also be fitted to the cultivator frames. They usually have adjustable wings and replaceable wear points. Two types are used, the potato type for soft soil and the Lister type for hard soil. There are also special purpose tie ridger machines used for erosion control in some areas: their use allows the water to form ponds to give time for percolation into the soil after a high-intensity storm.

7.1.3 Chisel plough

The chisel plough is designed to burst hard soil and leaves trash on the surface, and therefore it is more suited to areas where erosion may be a problem. The tines have adjustable lateral spacing to suit various row requirements. Many designs have replaceable wear tips and are protected by shear bolts as shown in Figure 7.2, a typical chisel plough.

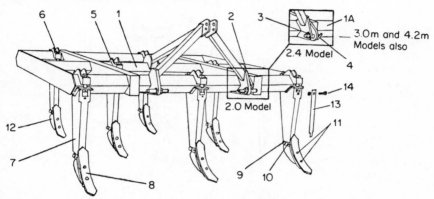

FIGURE 7.2 Chisel plough (Bomford and Evershed Ltd)

1. Frame 2.0 m	5. Clamp top	10. Shear bolt, nut, washer
1a. Frame 2.4 m and larger models	6. Clamp bolt and nut	11. Bolt and nut for point
2. Hitch pin nut, Washer	7. Standard tine only	12. Frog
3. Linch pin	8. Reversible point	13. Leg guard ⎫ if
4. Hand pin 2.6 m and larger models	9. Pivot bolt, nut, washer	14. Nut and bolt ⎬ required for leg guard ⎭

7.1.4 Rotary plough

The rotary plough is intended for wet ricefield cultivation. The plough consists of two rotors with spiral flights fitted on either side of the drive axle of a pedestrian single-axle tractor—see Figure 7.3.

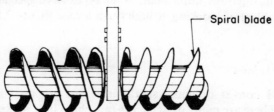

FIGURE 7.3 Rotary plough fitted to a single-axle tractor

7.2 SECONDARY TILLAGE EQUIPMENT

7.2.1 Disc harrows

Disc harrows are sets of discs on a shaft angled to the direction of travel, in sets of two or four and arranged so that there is no net side thrust on the tractor. The greater the angle, the greater the soil disturbance. They are used to cut up the clods from primary cultivation to make a seedbed. Larger sizes are angled hydraulically and have depth wheels; most types have weight trays in order to add weight for adequate soil penetration. Figure 7.4 shows a typical disc harrow.

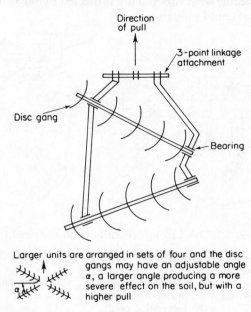

Larger units are arranged in sets of four and the disc gangs may have an adjustable angle α, a larger angle producing a more severe effect on the soil, but with a higher pull

FIGURE 7.4 Disc harrow

7.2.2 Cultivator

A cultivator is constructed like a light chisel plough but with more tines. It is used for stirring the soil to get uniform soil compaction with the fine material at the top, and some clod-breaking. Ridged and spring-loaded types are both available. The tines are usually set staggered on the frame to allow plenty of room for the trash to pass through the implement. Some types have specially angled legs to reduce the problem of blocking at high trash levels. Figure 7.5 shows a typical cultivator.

7.2.3 Light harrows

Light harrows consist of many small tines and are used for final seedbed preparation. There are many designs: zig-zag, chain link, spikes set in a wooden

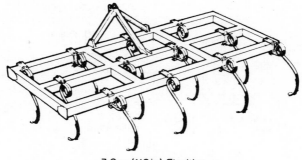

3.0 m (118 in) Flexitine

FIGURE 7.5 Cultivator—flexible-tine type (Bomford and Evershed Ltd)

pole, and so on. They are not suitable for working in conditions where there is a lot of trash. Tree tops or branches may also be used as harrows.

7.2.4 Rotary cultivators—powered

These machines have a horizontal shaft with various types of blades which cut and stir up the soil. The machines range from very small to very large and various types of blades are used: pick tine, spring-pick tine, rotary slasher blade, L blade, etc. These machines are well suited to work in rice paddy fields for stirring up the soil and incorporating trash. (The slasher type of blade should be used to prevent trash wrapping round the cultivator.) Apart from the use of a levelling board these are very often the only cultivation machine required. These types of machines are not suitable for hard soil conditions, except perhaps the very large heavy-duty types fitted to a big tractor, as they have severe vibration and wear problems under these conditions. Figure 7.6 shows the two main types in common use.

Powered vertical-axis types with tines are made only in the larger sizes and can be considered as special purpose machines for seedbed preparation for certain temperate crops.

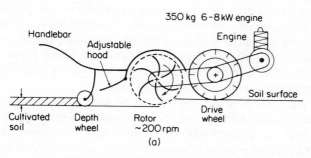

FIGURE 7.6a Power tiller (rotary tiller)

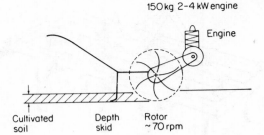

FIGURE 7.6b Motor tiller (rotor tiller)

7.2.5 Non-powered rotary cultivator

These machines can have horizontal and vertical axes and have blades or tines arranged so that they rotate as they are pulled through the soil. They are said to have a good stirring and mixing characteristic and to be resistant to blockage by trash. Some types are suitable for weed control in ridge cultivation systems.

7.3 SEEDERS

The objective of any seeder is to place the seed at the required depth and spacing in the soil at a suitable rate. They range from a simple hand-fed tube behind a scratch plough to a precision seeder planting single seeds from a belt at a high ground speed.

A seed drill controls the width of the row spacing and the depth of planting but has no definite control of spacing in the row, whereas a planter controls the spacing in the row as well.

Seeders may be divided into two major types, those required to plant through trash into undisturbed soil and those required to plant into loose bare soil. The former type, known as jab planters, may be single-shot, hand-operated types or multiple-spiked rotary types for hand or small tractor propulsion. The seed metering system is usually some form of roller with various pocket sizes to suit the seed being used, with a brush arranged to allow one seed per pocket to fall from the seed hopper into the injector point. The multiple-spiked types have a small lever which opens the seed spout below the ground surface at the correct depth, operated by the machine's own weight as it rolls along. Figure 7.7 shows a typical example. The minor disadvantage of this type of machine is that the distance of planting cannot be varied very easily.

Seeders to plant into bare soil may have a tine, shoe, or disc to open a groove or furrow of the required depth in the soil, followed by a metering mechanism to drop the seeds at the required rate. A tine or scraper covers over the seed and a press roller may follow to firm the soil over the seeds. The furrow opener should be resistant to being blocked by trash and should be designed so that the dry

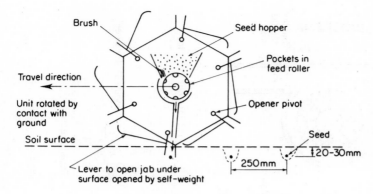

FIGURE 7.7 IITA multiple jab planter

surface soil does not fall into the bottom, otherwise the seeds will be planted in dry soil and not the damp soil just below the surface.

There are many types of metering mechanisms used, such as holes in cylinders or discs pushing seeds past a flexible barrier—brush or rubber flap: variations on this theme are very common. A cylindrical brush pushing seeds up to various sized holes is a well-tried type for small seeds, particularly those of vegetables and grass. Fluted rollers rotating near to an adjustable hole are widely used in grain drills. The feed rates are altered either by driving the mechanism faster in relation to ground speed or by increasing the hole size, or both. This can be done by using differing numbers of pegs on a cylinder and a lantern wheel, by an adjustable friction wheel and disc, or in the more elaborate types by means of gears, and different seed plates. Figure 7.8 shows some of the common seed metering mechanisms.

It is essential that the mechanism does not damage the seed and the hopper should be so designed that no bridging occurs, thus ensuring that the mechanism has an even flow of seed.

7.3.1 Calibration of planters and drills

It is important that the machine plants the seed at the required seed rate in kg/ha. Too much seed used is an expensive waste and not enough will probably lead to a poor crop stand and consequent loss of yield. The machine can be calibrated by holding it above the ground so that its metering wheel can be rotated. The seed is collected in a container and weighed after a certain number of revolutions corresponding to the distance travelled for say 1/10 ha. The number of revolutions required by the metering wheel to cover 1/10 ha is

$$N = \frac{10000}{10 \times x \times n \times \pi d}$$

where x is the row spacing in metres, n is the number of rows the planter can plant

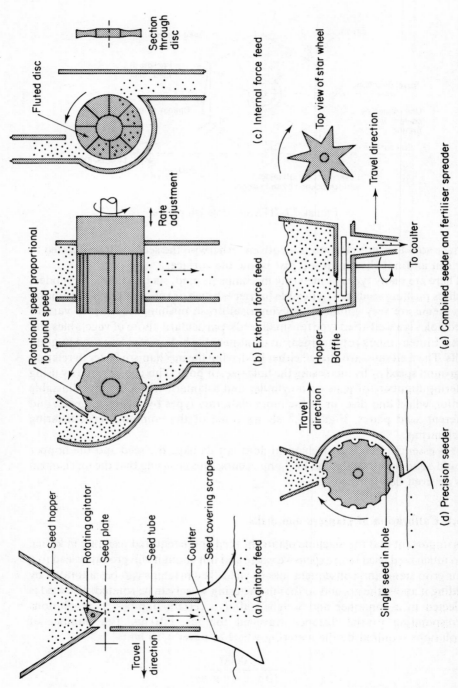

FIGURE 7.8 Feeder mechanisms

(a) Agitator feed

Seed hopper
Rotating agitator
Seed plate
Seed tube
Coulter
Seed covering scraper

Travel direction

Rotational speed proportional to ground speed

Rate adjustment

Fluted disc

Section through disc

(b) External force feed

(c) Internal force feed

Top view of star wheel

Travel direction

(d) Precision seeder

Single seed in hole

Travel direction

Hopper
Baffle
To coulter

Travel direction

(e) Combined seeder and fertiliser spreader

at a single pass and d is the rolling diameter of the metering wheel in metres. The weight of seed collected for N revolutions $\times$ 10 is then the seed rate in kg/ha.

There are also many specialized planting machines, for example the sugar cane/tapioca planter, the chitted potato planter, the rice transplanter, but these are not usually considered as suitable for the small farmer.

7.4 SPRAYERS

Sprayers are used to apply chemicals to the crop to control weeds, insects, and fungi, and also to apply plant growth regulators as a series of fine drops. Sprayers fall into two main categories: high-pressure hydraulic and spinning disc. The high-pressure hydraulic types use relatively large volumes of water to dilute the spray chemical which is pumped through a fan jet or swirl nozzle to form a fine spray. This technique requires a lot of water and relatively high powers and gives a spray with poor uniformity of drop size, with many particles too small to be of use which get lost as spray drift. Figure 7.9 shows the comparisons between the spinning disc and the hydraulic nozzle type of applicator.

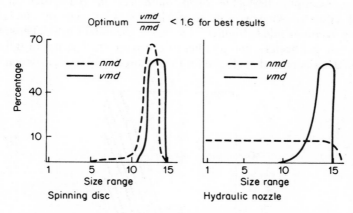

FIGURE 7.9 Comparison of spinning disc and hydraulic nozzle

The volume median diameter (vmd) of the drops is a measure of the diameter of the drop such that the drops of a larger diameter contain half the volume of the total number of drops. The number mean diameter (nmd) of the drops is the diameter such that half the number of drops in the total number is larger and half smaller. The ratio of vmd/nmd is called the dispersion ratio. A hydraulic sprayer commonly has a ratio of 3 : 6 whereas a spinning disc sprayer has a ratio around 1 : 6. A low ratio shows that there is much better control of the drop size, which is usually more efficient in terms of the quantity of chemical required and the volume of water required (degree of dilution) for a particular application.

The spinning disc type is becoming widely used as much smaller spray volumes are used but with much better drop size control. These types are commonly called

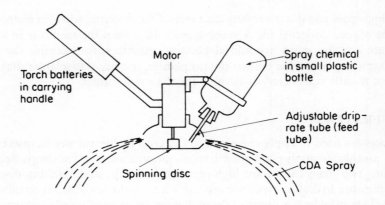

FIGURE 7.10 CDA spinning disc sprayer

CDA, controlled droplet application (see Figure 7.10). The volume of chemical required per hectare is related to the drop size required for the particular purpose. For example, herbicide drop diameters in the order of $300 \mu m$ are most effective whereas for insecticides a drop size of $20-120 \mu m$ is commonly recommended. Figure 7.11 shows the relationship between drop diameter, rate per hectare, and number of drops/sq cm. The size of the disc and number of discs can then be selected for the required output. A particular disc will

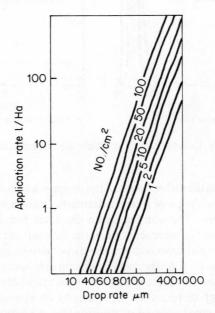

FIGURE 7.11 Relationship between drop diameter, rate per hectare, and number of drops per square metre

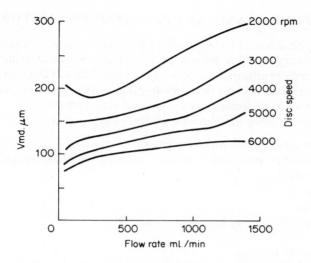

FIGURE 7.12 Effect of flow rate and disc speed on
volume median diameter (vmd)

have a characteristic similar to that shown in Figure 7.12 so that the disc speed, drop diameter, and flow rate can be determined. From the flow rate and the number of units selected the ground speed can be determined.

A technique of drift spraying is commonly adopted by spraying across the wind direction; since the particles are of the correct size they will all settle out on the crop and very little will be lost. The chemicals used are usually in an oil-based material so that losses due to evaporation are minimal. Small hand-held machines can spray bouts of a few metres wide whereas larger multiple-disc types can produce up to 100 metre swath widths. There is little risk to the operator from exposure to the spray chemicals if he adopts the manufacturers' recommendations. The smaller machines are driven by small electric motors using torch batteries in the carrying handle. Motor speed is controlled by varying the voltage, that is by varying the number of batteries fitted. A simple vibrating wire speed indicator and vane or a float-in-tube wind speed indicator are useful aids to better accuracy in the field.

The sprayer can be calibrated by calculating how long it takes to walk or drive over say 1/10 hectare at the particular width that the sprayer can cover and measuring the quantity of fluid that is dispensed in this time (water can be used for calibration purposes).

It is essential to follow the instructions supplied by the spray chemical manufacturers regarding the quantities to be used and the safety precautions that are necessary. Many of the chemicals are extremely poisonous and must be stored in a secure place. All empty containers must either only be reused for the same chemical or else destroyed at once.

7.5 ROOT ZONE GRANULAR FERTILIZER APPLICATORS

Root zone application of fertilizer offers good potential for increasing the net income of farmers by either reducing the quantities required or increasing the yields through more effective use. Several granular chemical fertilizer applicators have been developed, notably in the Philippines, with a view to encouraging the adoption of root zone technology by lowland rice farmers.

The various metering mechanisms which can be used are:

(a) an adjustable orifice without agitator;
(b) an adjustable orifice with agitator (very high application rate);
(c) a star wheel;
(d) a chain;
(e) an auger;
(f) a fluted roller (fairly high application rate);
(g) a belt;
(h) a revolving bottom;
(i) an ascending bottom;
(j) a vertical rotor with pockets;
(k) a plain roller (high application rate).

7.6 MECHANICAL WEEDERS

Mechanical weed control is achieved by disturbing the soil and/or cutting the roots so that the weeds die from root disturbance or dessication. Obviously it is important to ensure that weeds are at the right stage of growth and that soil and climatic conditions are suitable as this will affect the efficiency of the operation considerably. (Chemical control must also be done at the right time taking similar factors into account.) Many designs of cutting and stirring implements are used, A-shaped blades between the rows and L-shaped blades close to the plant. The depth of work should be adjusted to disturb the roots of the weeds but not those of the crop. Crops planted in rows are easier to hoe, and if a tool-bar-mounted weeder is to be used it should be set at the same row spacing as the original planter or many of the crop plants will be lost during the weeding operation. Weeding is required only at the early stages of the crop until it gives a complete ground cover which then discourages weed growth naturally.

7.7 HARVESTING EQUIPMENT

7.7.1 Grass cutting

For grass as a weed or decorative ground cover a vertical-axis rotating blade or disc with peripheral blades can be used. Machines are made both in small and large sizes with a tip speed in the order of 80 m/s. Strong guards must be fitted to protect the operator from the blades and from flying stones. For grass as a crop a reciprocating knife design is used for low-powered small machines while a vertical-axis drum with swinging blades (drum mower) can be fitted to a tractor and operated at 10–16 km/hr forward speed.

For grass to be stored as silage a haylage or forage harvester can be used, but flail types and double and single chop types are designed for large high-powered machines and are not made in small sizes.

7.7.2 Root crops

Various sweep blades and shares are used to loosen or push the crop to the surface to enable hand picking. More elaborate machines have an elevator chain or spinning rotor behind the share to leave the crop on top of the soil surface. Elevator diggers can be fitted with additional agitators on the chain for better soil separation and even larger, more elaborate machines elevate into a sorter and then into a trailer.

7.7.3 Grain crops

Many types of knife and sickle are used by hand depending on local tradition. Aids fitted to scythes make it easier to cut bundles to be hand tied. Animal-pulled reciprocating cutters for hand-tie bundles can be improved upon by fitting self-tying mechanisms as in a simple reaper and binder. The bundles may then be dried and stored or taken to a static field thrasher powered by a small engine. The mechanisms used vary considerably but a common type consists of a horizontal-axis drum with bars or wire loops rotating at a peripheral speed ranging from 10 to 30 m/s depending on crop type. (See Figure 2.8 for an illustration of a small thrasher.) Arranged close to the drum is a concave also consisting of bars or loops with a nominal clearance of 12–50 mm depending on the crop. The width of the drum is from about 0.3–1.5 m and it is powered by an engine of a few kilowatts output. The unit can be carried by two to four men from field to field. In some types just the heads are held into the rotor while the higher output types take the whole crop through the machine. The straw and grain are usually separated manually although simple fans and sieves are sometimes used. By suitably adjusting the rotor speed and concave clearance other crops such as peas, beans, and lentils can also be thrashed. The machine should be adjusted for a particular crop according to the manufacturer's instructions but in general the operator should use the largest clearance at the slowest speeds consistent with good thrashing. This will minimize damage to the crop and use least power.

Self-propelled harvesting machines such as small combine harvesters are not widely used by the smaller farmer as the capital cost is rather high. Where the economic conditions are suitable, however, small tracked combines can be used very successfully, for instance for rice in Japan.

Many attempts have been made to make a stripper type harvester which just takes the grains from the standing crop. The advantage of this would be a machine with a single rotating component needing only a few simple adjustments, but to date there is no effective solution to the high shed losses and the inability to deal with laid crops.

7.8 OTHER TASKS

Tasks considered are: erosion control works, levelling paddy fields, drainage ditches, site preparation for buildings, and so on. Various types of angled plates can be used with manual control of cutting depth for grade levelling. The larger type of small tractor can be fitted with hydraulically operated shovels and digger buckets which can carry out most tasks. The larger earth-moving jobs are best carried out by large gangs of men or contractors using specialized equipment.

7.9 TRANSPORT

Skids, transport boxes, trailers and trucks of various designs are used. See Chapter 3 for a discussion of transport requirements.

7.10 MOUNTING OF IMPLEMENTS

The method adopted to mount the implement to the prime mover has a great bearing on the overall efficiency of both. The basic requirement is that the implement should be attached in such a way that the forces between the implement and the prime mover are carried most effectively.

The following discussion applies to forces in the vertical plane; forces in the horizontal plane which affect steering have been dealt with in Chapter 6. The situation is considered quasi-statically, that is with the implement moving at constant speed in uniform conditions. Power for acceleration is not usually considered for agricultural machinery and design for shock loads will not be considered here.

The forces present for a soil-engaging implement are as shown in Figure 7.13. All the forces can be estimated, for example weight, soil force (see later) or calculated from the force diagram, for example depth wheel force. If Figure 7.13 is drawn to a suitable scale and the forces also drawn to some scale the values of each can easily be calculated. If the drawbar is moved downwards at the implement end, the force on the depth wheel is reduced giving less rolling resistance loss, and the draught force has a larger vertical component which will improve the traction characteristics of the tractor (see Chapters 6 and 9). See the dotted lines on Figure 7.13 to illustrate this point. It will be noticed that the

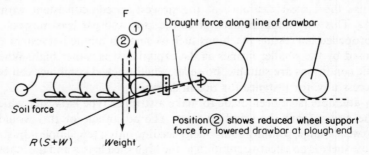

FIGURE 7.13 Forces on a trailed implement

effective depth wheel support force becomes smaller and moves further back. The actual force supported under each wheel can then be calculated if the distances between the wheels are known (see Figure 7.14). The rolling resistance of the depth wheels can be allowed for by drawing the support force at a backwards angle, the tangent of the angle being the coefficient of rolling resistance.

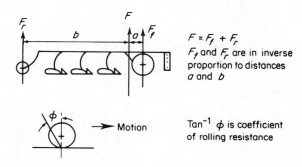

$$F = F_f + F_r$$

F_f and F_r are in inverse proportion to distances a and b

$\mathrm{Tan}^{-1}\,\phi$ is coefficient of rolling resistance

FIGURE 7.14 Support forces on trailer implement wheels

Instead of a simple drawbar the tractor may be fitted with a three-point linkage, a typical form of which is shown in Figure 7.15. The implement and the tractor can then be made to operate as a single unit. The two lower links take the main draught force, and the top link length is adjustable to control the pitch of the implement. The check chains go tight when the implement is lifted to prevent sideways swing in the transport position. The stabilizer bars control side movement in work and are used for above-soil implements such as grass cutters. The lift rods can have a number of hole positions to adjust the maximum or minimum lift height positions. The right-hand rod has a levelling box so that the implement can be set level or tilted (particularly for mouldboard ploughs). The top link may have a number of positions which will affect the draught control signal into the top link spring and also affect the maximum lift capacity of the linkage. The linkage should be nearly parallel for maximum lift capacity. Only large tractors are fitted with bottom link draught control or with a torque-sensing transmission linkage control. All the types are operated by an engine-driven hydraulic pump and lift cylinder. The linkage may be depth-wheel controlled. Figure 7.16 shows the lines of force arising from the soil and weight forces on the implement. The ICR is the instantaneous centre of rotation of the links and is the imaginary point of convergence—usually near the front of the engine. The virtual hitch point is the effective hitch point, ie where a chain could be fixed to pull the implement, and is where the forces' centre acts.

If a depth wheel is not fitted the forces can be balanced by moving the bottom link up at the tractor end. This moves the ICR and VHP to position (2). (This system is used on some small tractors.) The bottom link can be moved down at the tractor and to make the implement work deeper. If neither a depth wheel nor an

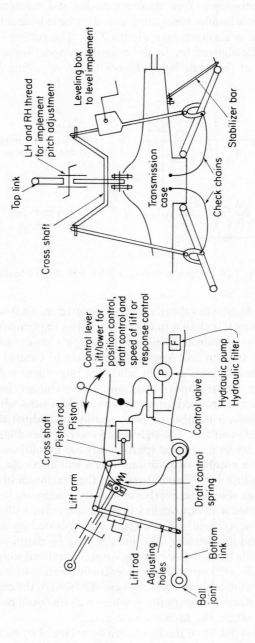

FIGURE 7.15 A typical tractor three-point linkage

Top link

LH and RH thread for implement pitch adjustment

Leveling box to level implement

Cross shaft

Transmission case

Stabilizer bar

Check chains

Control lever Lift/lower for position control, draft control and speed of lift or response control

Cross shaft

Piston rod

Piston

Control valve

Hydraulic pump

Hydraulic filter

Lift arm

Draft control spring

Lift rod

Adjusting holes

Bottom link

Ball joint

F

P

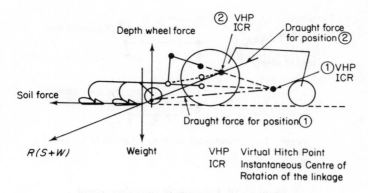

FIGURE 7.16 Forces in a three-point linkage

adjustable linkage is fitted a draught control system can be used which gives a force in the lift rod. Figure 7.17 shows the forces present in this system. The draught control works by sensing the force in the top link using a single-acting spring (smaller tractors) or a double-acting spring (medium tractors). See Figure 7.18 for an illustration of this. The hydraulic control valve has three positions, Lower L, Neutral N, and Raise R which direct the hydraulic oil into or out of the lift cylinder. An increased draught force causes the top link sensing arrangement to put the control valve into the raise position, lifting the implement so as to reduce the draught force to the present value. A reduction in draught force causes the opposite effect.

Many tractors are also fitted with a response control which affects either the rate of lift or the rate of lower of the implement while the draught control system is controlling the implement. This avoids the system oscillating vertically, particularly for long soil-engaging implements operated at high speed.

Above-ground operating implements may have their height controlled by the position control—see Figure 7.19. Movement of the lever downwards shortens the rod so that the implement goes nearer to the ground; if it overshoots the valve

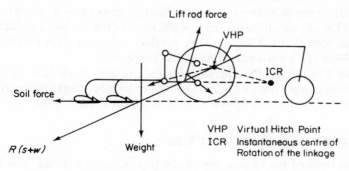

FIGURE 7.17 Forces in a draught control system

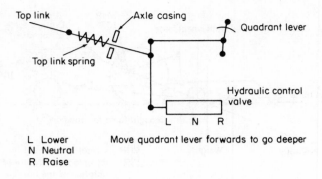

Top link Axle casing Quadrant lever

Top link spring

Hydraulic control valve

L N R

L Lower
N Neutral
R Raise

Move quadrant lever forwards to go deeper

FIGURE 7.18 Principle of draught control

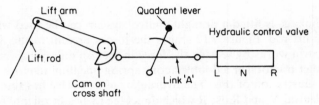

Lift arm Quadrant lever

Hydraulic control valve

Lift rod

Cam on
cross shaft

Link 'A'

L N R

Quadrant lever moved forwards shortens Link 'A' and vice versa

FIGURE 7.19 Principle of position control

goes into raise and the implement is lifted slightly. Other control methods are also used, such as Traction Control Unit and pressure control. Both systems employ an adjustable pressure regulator in the system so that a constant lifting effort can be applied by the lift rods to give the required effect of weight transfer.

As can be seen from Figure 7.17 the line of action of the draught force is steeper than in Figure 7.16. The steeper the line of the draught force the greater the load on the rear wheels, with correspondingly less on the front. The implement should be matched carefully to the tractor, with the maximum front weights fitted and rear weights on the implement so that the draught line of action can be as steep as possible. This will maximize the tractive efficiency of the tractor. Too high a force impairs the steerability (Chapter 6) or overloads the rear tyres and thus sets a practical limit for a particular tractor and implement.

The power to operate the hydraulic lift and draught control can be 6–10 kW, and therefore smaller tractors have to use depth wheels or linkage control due to their power limitations. A simple up and down lift is then fitted.

7.11 THE SINGLE-AXLE TRACTOR

The interactions of the various forces can be studied in a similar way by using scaled dimensions and force diagrams. The aim of the adjustments should be to make the plough run at the required depth without any need for the operator to supply a hand force continuously. The centre of gravity should be as close to the wheel as possible so that the wheel force is nearer vertical in position (2) to give better traction characteristics. A downward hand force moves the wheel force to (3), thus increasing the wheel thrust required. The screw adjustment which changes the angle of the implement moves the centre of gravity back and enables the implement to work deeper. Figure 7.20 shows the positions of the various forces referred to.

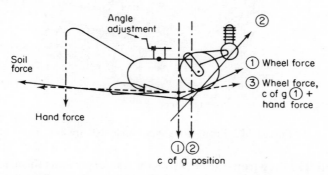

FIGURE 7.20 Forces in a single-axle tractor with soil-engaging implements

Some implements have a downward component of the soil force, for example a blade or plough in a soft soil condition. In this case the centre of gravity can be in front of the wheels and its position adjusted by the front weight being moved forwards (see Figure 7.21). When the operator lifts the handles to turn round, the machine lifts easily and he has to provide a downward hand force to balance it—this is convenient for the operator in the difficult working conditions which prevail in soft soil. Light draught implements should be fitted as close to the wheels as possible to give weight near to the drive wheels and to give the operator maximum foot space for walking behind.

Another single-axle design is the front-drive type as shown in Figure 7.22. The puddling wheel or rotor is driven by the engine and produces a forward thrust

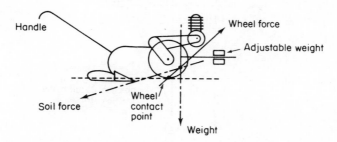

FIGURE 7.21 Single-axle tractor with front weight

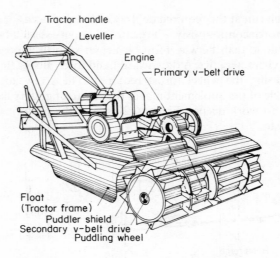

Tractor handle
Leveller
Engine
Primary v-belt drive
Float
(Tractor frame)
Puddler shield
Secondary v-belt drive
Puddling wheel

FIGURE 7.22 Front drive type single-axle tractor

from the soil. The drag from the float (tractor frame) prevents the machine from travelling forward at the same speed as the peripheral speed of the rotor. The rotor thus behaves as a slipping wheel causing the required amount of soil disturbance.

The single-axle tractor may be converted into a powered trailer as shown in Figure 7.23. From the figure it can be seen that the nearer the combined centre of gravity is to the drive wheel the better the traction performance, as the resultant support force is more vertical. A wide, close-coupled trailer with wheels at the back will be much better than a narrow trailer with its wheels under the pay load.

Various types of trailers may be fitted to conventional tractors. Figure 7.24 shows the two basic types. A balanced trailer will not add much weight to the tractor drive wheels whereas the trailer with wheels near the back will on-load the

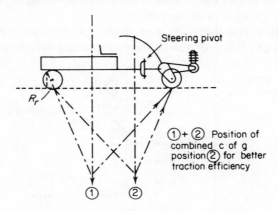

Steering pivot

R_r

① + ② Position of combined c of g
position ② for better traction efficiency

① ②

FIGURE 7.23 Single-axle tractor and trailer

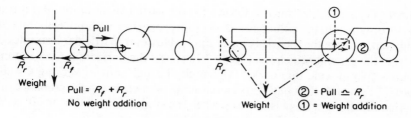

FIGURE 7.24 Trailer attached to conventional tractor

rear of the tractor considerably, thus helping traction in poor conditions. A pick-up hook will however be necessary for this type of trailer, due to the large vertical drawbar load.

7.12 THE WINCH-PULLED IMPLEMENT

The winch-pulled implement shown in Figure 7.25 has the resultant force of the soil and weight intersecting the rope force at (1) which gives the magnitude and position of the support force required. If the soil force increases then the support force changes to position (2). The force can be on a wheel or skid. If on a wheel then it must be adjusted both longitudinally and vertically for a change in working depth, whereas a skid needs only to be adjusted vertically for a depth change provided it is long enough for the range of forces encountered.

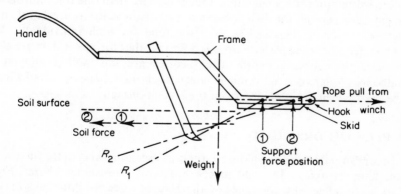

FIGURE 7.25 Winch-pulled implements

7.13 PULL AND SPEED REQUIREMENTS OF IMPLEMENTS

Most primary cultivation requires a high force to be transmitted from the implement to the soil to give the required effect. There are a number of theories which can be used to predict these forces from a knowledge of the physical properties of the soil and the geometry of the soil-engaging part of the implement.

The basic soil parameters of cohesion and friction (see Chapter 6) and a

166

number of other factors must be known for the particular situation. But it is not easy to obtain reliable values for these soil properties without elaborate test equipment, nor are the theories easy to apply. However, particular examples can illustrate the general principles. Figure 7.26 shows the horizontal force required to pull a 50 mm wide tine of 90° rake angle through soils of varying C and ϕ values. The general effect of the rake angle is also shown. From the graph it can be seen that the draught force is doubled when the angle is changed from 30° to 90°.

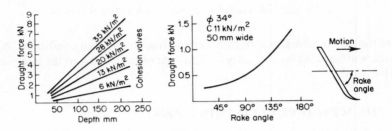

FIGURE 7.26 The effect of depth, rake angle, and cohesion on draught force

The vertical forces also change; a 30° raked tine will tend to sink in the soil whereas a 90° tine will require a vertical force to hold it down to the working depth. As the tine is probably expected to work at a constant depth this will affect the method of depth control which in turn will affect the prime mover. Figure 7.26 shows the situation for a single tine; if there is more than one tine and they are mounted close enough for their spheres of soil disturbance to overlap then the draught of a set of tines is less than the total for each individually. Very approximately the force for two tines is 1.5 times that for one if they are closer than twice the working depth. The theories are less well developed for mouldboard ploughs due to the much more complicated shapes involved. Graphs can be used which are obtained from the results of many field tests.

7.14 PLOUGH DRAUGHT

The force required to pull any plough is the cross-sectional area of the furrow slice × ploughing resistance. Draught in kN = area m² × resistance kN/m². Figure 7.27 shows the effect of travel speed on the draught force, for different soil types. In general higher speeds cause higher wear and damage due to shock loads, particularly in conditions when rocks and tree roots are present. For other implements, typical values of pull and speed for average conditions can be tabulated from records of practical field tests.

7.15 MACHINERY PERFORMANCE DATA

Table 7.1 shows performance data for some commonly used machines. It should be noted that these are typical figures under average conditions and that actual

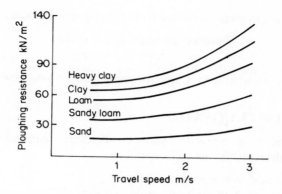

FIGURE 7.27 Ploughing resistance against travel speed for various soil types

figures under particular conditions can vary very widely. For instance in a dry year in heavy clay soil before the rains draught forces can be two to three times those stated, whereas in a wet year the limitation will be tractor wheel slip and sinkage. Both these problems can be overcome by the appropriate action (see Chapter 6 for Traction Aids). Table 7.1 also shows a factor called field efficiency. This factor allows for the reduction in performance due to poor bout matching and wasted time turning at the headlands.

The work output of the implement may now be calculated:

$$\text{Ha/h} = \text{width} \times \text{speed} \times \eta \times \frac{3600}{10000} \text{ (width in m; speed m/s)}$$

TABLE 7.1 Implement Performance

Machine	Draught kN	Speed m/s	Field efficiency %
Mouldboard plough	1.4–3.5	1.2–2.5	70–80
Lister ridger	1.8–3.6	1.3–2.5	70–90
Subsoiler: Light soil	11–20 N/mm depth	1.3–2.2	70–90
Heavy soil	20–30 N/mm depth		
Rotary cultivator	10–30 kW/m width	0.5–1.5	
Disc harrow	1.5–4 per m width	1.3–2.2	70–90
HD cultivator: Light	2.2–7.5/m width	1.3–3.6	70–90
Heavy	5–10/m width		
Light cultivator: Light	0.6–1.1 kN/m width	1.1–2.2	70–90
Heavy	11–22 N/m width/ mm depth	0.7–1.3	
Grain drill	0.4–1.5 kN/m width	1.0–2.7	65–85
Mower: reciprocating	2.5 kW/m width cut	2.2–3.0	75–85
rotary blade	7.4–20 kW/m width cut	1.3–3.6	75–85
Small grain combine	30 kW/m width drum	0.9–1.8	65–80

and the power requirement calculated from the required or recommended speed.

$$\text{Drawbar power} = \text{pull} \times \text{speed}$$
$$\text{kW} = \text{kN} \times \text{m/s}$$

Note: this is the drawbar power, *not* the engine power (see Chapter 6).

7.16 DEPTH OF CULTIVATION

It is very difficult to give precise figures on the depth required of the various cultivation implements as this will depend on:

(a) factors related to the machine; and
(b) factors related to the soil and crop.

Machine factors include: strength limitation resulting in depth of work limitation; function limitation, for example mouldboard ploughs will not work at a depth less than a quarter of the width, or they do not turn a proper furrow; crop clearance limitation; speed—too high a speed may cause bouncing, wear, and impact damage under poor conditions.

Soil factors include: the cultivation should leave the soil in a suitable state for crop establishment and growth; small seeds require a fine seedbed to give good soil contact and correct depth of soil cover, but not too high a compaction that will limit the free movement of air and water in the soil; a cultivation depth sufficient to give break-up of an impervious pan; perhaps subsoiling every third season. In general the depth of work should be as shallow as possible consistent with the crop requirements as the draught is then likely to be a minimum which in turn will allow faster working for a given power availability. This can lead to better timing of operations and probably better agricultural performance in terms of crop yields.

Part 3 *Matching the Equipment to the Task*

8 *The Economics of Small Farm Mechanization*

J. Morris

Animal power in the right context can be an effective means of mechanization

8.1 INTRODUCTION

Almost without exception, Third World governments emphasize the role of agriculture within their overall development strategy, with particular reference to increasing food production, rural incomes, and employment opportunities. With the smallholder/peasant sector often accounting for over 70 per cent of total agricultural production and over 50 per cent of the economically active population, 'Green Revolution' programmes are necessarily small-farmer oriented. The overall purpose of such programmes is to improve small farm productivity, and further mechanization is one possible way of achieving this. However, a number of features of the smallholder system and of mechanization technology itself are important in influencing the scope for mechanization and its potential contribution. This chapter, in very general terms, considers the basic features of smallholder systems relevant to mechanization, and the nature of mechanization and its potential benefits. The need and justification for mechanization on the small farm is also discussed. The chapter also contains a comparative analysis of alternative smallholder mechanization systems in terms of selected criteria, and concludes with an overview of small farm mechanization policy.

8.2 FEATURES OF THE SMALLHOLDER SYSTEM RELEVANT TO MECHANIZATION

It is now largely agreed that the new and improved technologies to be adopted by developing countries must be both appropriate and acceptable, not only in terms of technical suitability, but also with particular regard to the resources and aspirations of their recipients. In this context, the mechanization planner needs to consider a number of features of the 'smallholder target group' which influence the need and scope for mechanization. These can be generalized as follows:

(1) *Farm Size and Structure.* Some 80–90 per cent of holdings in developing countries are below 5 hectares, and often 50–60 per cent are 2 hectares or less. Land tenure arrangements which encompass the rights to own and/or use land are notoriously varied and complex. Owner-occupancy, tenancies with varying degrees of security and rental agreements, and communal ownership which grants usufructory (land use) rights, have a differential effect on the farmer's ability or willingness to invest in his farming system. Limited access to land ownership, for instance, often means that farmers do not possess the collateral needed to qualify for medium-term machinery credit. Inheritance customs, combined with population pressure, can lead to excessive fragmentation and dispersal of holdings. Smallholder land itself can further limit mechanization feasibility; namely, topography, drainage, natural vegetation, and accessibility. The overall effect is to reduce the scope for sophisticated farm power systems more suited to large contiguous machinery-oriented holdings.

(2) *Population and Labour.* Small farm systems are by definition labour-intensive and family-oriented. In many cases the smallholder sector engages over 50 per cent of the economically active population. Family sizes generally average six to eight persons, usually with two or three adult male-equivalent workers (just over one man per hectare). Family labour is usually the most abundant resource, and whilst it may remain under-employed for much of the year, shortages during critical periods, such as weeding, can act as a constraint on overall productivity. Larger than average farms are important sources of paid employment. With populations increasing in excess of $2\frac{1}{2}$ per cent per year, and the lack of employment opportunities in non-farm sectors, many countries look to the smallholder sector to absorb rural population growth.

(3) *Semi-subsistence Farming.* The average 2–3 hectare holding generally devotes 60–70 per cent of its area to household food crops, mainly cereals (at yields of 1 t/ha, a family of six would need about 1.6 hectares cereal

equivalent to sustain itself). Crop and livestock production are typified by low and unreliable yields, unimproved species, low use of fertilizer or animal fodder, and limited pest and disease measures. The more intensive smallholder systems involve dryland intercropping, or irrigated relay cropping.

(4) *Low and Variable Incomes.* All the foregoing features combine with low produce prices to generate low and uncertain disposable incomes for smallholders. The arithmetic is simple; at world market commodity prices a 2-hectare holding producing say 1.5 tonnes of cereals at £80 and 0.5 tonnes of groundnuts at £250 is equivalent to about £250 gross income for six people. Most farmers receive much less than world market prices. Even with improved practices the situation is not much different; a 2-hectare holding in northern Nigeria (1978), Egypt (1979) and Malawi (1980) was estimated by the author to give net incomes of about £500, £350, and £200 respectively in 1981 prices. This level of income leaves little room for any form of new technology, much of which remains expensive and inherently risky.

(5) *Institutional Support.* Input supply, marketing, credit, extension, and training are limited and a major constraint to improved productivity. In the past, such services have been aimed at the larger, more progressive farmers. Very recently, a number of countries, for example Malawi and Nigeria, have devised more modest and appropriate 'basic service' packages which provide input supply, marketing, credit, and extension. Selective mechanization has featured through medium-term credit provisions and training schemes, mainly related to the use of oxen.

8.3 FARM MECHANIZATION AND RELATED BENEFITS

In its broadest sense farm mechanization is to do with implements, machines, and power sources. In economic terms mechanization involves injecting extra capital into the farming system mainly with a view to increasing labour's capacity to do work defined in terms of quantity and/or quality of output per worker. The potential benefits of mechanization to the farmer are reduced drudgery, increased returns, and reduced costs.

Reducing drudgery is important in retaining a commitment to farming by the young, and in releasing women and children particularly from the toil and tedium of many farming operations to spend their time more productively on other activities.

Mechanization can generate increased returns which manifest themselves in a variety of interrelated ways:

(1) Increased yields per hectare due to improved timeliness of operation and an improvement in the quality or precision of the task. Mechanization can contribute indirectly by facilitating the application of yield-increasing inputs such as chemical fertilizers and irrigation water.

(2) An extension of the cultivated area made possible by the capacity to do more work in the time available.

(3) New crop and livestock systems which were previously impracticable or for which labour was not available. Refrigerated milk collection points are often a prerequisite for smallholder milk projects, just as tractor-hire services for primary cultivation may be required to ensure the feasibility of double-cropping irrigation systems. Furthermore, mechanization of routine operations on staple crops could release labour for more labour-intensive, high-value cash cropping.

(4) Higher commodity prices arising from a more reliable and regular quantity and quality of supply. Investments in storage and transport facilities can reduce temporal and spatial price uncertainties and increase the share of final commodity prices accruing to the producer.

(5) Generation of off-farm employment opportunities in sectors servicing agricultural mechanization (manufacturers, dealers, repair workshops).

Where mechanization is substituted for hired farm labour, cost savings may be apparent to the employer. In some situations, where labour is unreliable or difficult to manage, farmers may prefer mechanization, even if it costs more to use machinery.

An additional benefit to the private farmer is the prestige associated with machinery ownership and use. Like many 'modernizing' investments, spending on mechanization often includes a 'consumption' as well as a 'production' component; an outward sign of progressive thinking; the considerable inventories of unused but cherished oxen equipment bear witness to this.

In the main, mechanization is a 'labour-augmenting' technology increasing output per worker rather than output per unit of land. The benefits of mechanization have been greatest where labour is scarce (and therefore expensive) and/or land is plentiful. This characteristic of mechanization has important implications for its role and impact in the smallholder system, where (and generalization can be dangerous), for the most part land, capital, and management are limited, and labour is generally abundant.

8.4 ALTERNATIVE MECHANIZATION SYSTEMS

Mechanization systems are often categorized into man, animal, and engine-powered technology on the basis of sophistication, capacity to do work, costs, and in some cases precision and effectiveness. An indication of the relative importance of these power sources is given in Table 8.1 where the geographical variation reflects environmental and historical factors and the stage of agricultural development (Giles, 1975).

The predominant form of smallholder technology is that based on manual labour, with the hand hoe as a basic ingredient. The main attributes of this system are that it represents a low-cost, low-energy, labour-using, family-oriented technology, which is closely attuned to traditional farming methods such as

TABLE 8.1 Agricultural Power by Source and Geographical Region

Region	Total kW/ha	% of available power/ha		
		Man	Animal	Engine
Asia	0.16	26	51	23
Africa	0.08	35	7	58
Latin America	0.19	9	20	71
Total %		24	26	50

Source: Giles (1975).

minimum tillage and intercropping, and is largely self-sufficient, drawing on locally made implements. Furthermore, hired employment with payments in cash or kind is an important source of off-farm income. The main disadvantages of human-powered technology are that it often requires long, hard, and tedious work and the low level of labour productivity acts as a constraint on output and incomes. However, because of the primitive nature of many traditional implements, much scope exists for increasing labour productivity by improved hand tools and man-powered machines. Examples are the replacement of knives with scythes, and the introduction of hand-operated maize shellers.

In some areas work animals such as donkeys, oxen, cows, buffaloes, mules, horses, and camels are a common feature for farm transport if not for field work. They are especially common where there is a long history of animal husbandry, where tsetse fly is not a problem, and where farms are relatively large (above 3 hectares), population pressure is low, and land is available for grazing or fodder production. The advantages of work animals are that they can offer a relatively low-cost, low-energy, self-supporting, reproducible (with the possible exception of mules), and potentially comprehensive system of appropriate mechanization. Animals can improve labour productivity and help overcome major power constraints for small farmers without displacing labour. Animals can also provide a basis for informal contract hire. The main disadvantages of animal-powered technology are that it requires animal-husbandry skills which are not always in evidence, and animals and equipment remain expensive for the average smallholder. Furthermore, animals have a limited power capability, particularly if underfed, and feeding costs can be high especially where fodder and grazing land are limited. Work animals also require their share of institutional support, particularly regarding the supply of suitable animals and equipment, credit, training, and veterinary services. Animal draught power is often seen as the most 'appropriate' mechanization package for smallholders. The critical factors determining uptake are usually the availability (and cost) of oxen, feed, and the acquisition of animal management skills. Oxenization packages include oxen, a range of equipment, credit provisions, and operator and animal training.

Farm size and income largely preclude the average smallholder from acquiring engine-powered technology solely for use on his own farm. In the main, it is not possible to scale engine-powered technology down to the level where it is technically or financially suited to the individual smallholder. Stationary power units driving processing machines have a particularly important role, as do tractors for land preparation and transport but their potential is in terms of multifarm use through co-operatives, private contractors, or government hire schemes. In this context, the advantages of engine power are that it undoubtedly makes the greatest contribution to reducing drudgery. High-power, high-capacity, tractor-based systems theoretically offer the greatest achievement of the previously enumerated mechanization benefits, particularly those resulting from improved timeliness, new cropping patterns, and an extension of the cropped area. Engine-powered systems can encourage general farming modernization, and concomitant benefits such as the acquisition of new mechanical and management skills. The disadvantages of engine-powered systems are well documented. They are relatively expensive to acquire, operate, and maintain (in spite of often heavy subsidies), require a high (largely non-renewable) energy input, and often represent a non-indigenous, high foreign exchange technology. Multifarm systems are often difficult to organize and manage, and usually need considerable institutional support (Lonnemark, 1967). Engine-powered systems have been particularly criticized for their undesirable social and environmental impact, especially the displacement of labour in conditions of general under-employment. Small tractors (below 15 kW), however, have been designed with the smallholder farms in mind, but except for wet-land power tillers, they have met with little success, mainly because to date they have proved expensive to buy and operate, have demonstrated high costs per unit of work, and have generally not been powerful or heavy enough to perform primary tillage satisfactorily. In addition they require an order of institutional support similar to the conventional tractor (Pollard and Morris, 1978).

8.5 THE MECHANIZATION NEEDS OF THE SMALL FARM

The mechanization needs of the small farmer will vary according to the power requirements of his farm (as determined by the farm size and production system) and the extent to which existing power supply is a constraint on improving output. From the farmer's viewpoint, the justification for acquiring more mechanization will depend on such factors as:

(1) The financial worthwhileness of the mechanization investment; how the benefits compare with the costs.
(2) The ability to finance the proposed investment.
(3) The opportunity cost of the mechanization investment; would it be better spent on other things?

Where the choice exists, farmers will weigh up the relative attributes of alternative approaches to mechanization. Although not expressed in such terms, power performance, work capacity, cost effectiveness, and financing capability are likely to be important selection criteria. At the same time, national governments will be interested in the social, economic, and political ramifications of alternative approaches to mechanization, and they will be concerned with encouraging individual farmers, by a variety of policy measures, to adopt mechanization systems which are in the overall national interest. Some of these criteria are examined in turn.

8.6 CRITERIA FOR SELECTING MECHANIZATION

8.6.1 Power

Power per cultivated hectare can be used as an indicator of existing mechanization, and as a basis for mechanization planning. Studies of the relationship between effective kilowatts per hectare and average aggregate yields for major crops show that traditional smallholder systems have about 0.1 kW/ha. This reflects the observation that one man provides around 0.07 kW and that there are often one to two workers per hectare.

Considerable improvements in smallholder performance can be achieved mainly by the use of improved inputs (other than mechanization) which increase yields per hectare, such as improved seed, fertilizer, pest controls, and irrigation, and these can be facilitated by modest increases in power inputs. For instance, traditional cereal yields of say 0.8 t/ha can be increased to about 2.5 t/ha for an increase in power input of 0.3 kW/ha (Giles, 1975). Note 1 tonne (t) is 1000 kg.

In terms of power sources (Table 8.2) a supplement of 0.3 kW/ha would require another four men per hectare using traditional hand methods. Alternatively, a pair of well-fed 500 kg oxen can pull 10–20 per cent of their weight and exert a draught force of about 100 kg; enough to plough one furrow to

TABLE 8.2 Work Potential of Alternative Farm Power Sources

	Man (1 man)	Animal (1 pair of oxen)	Tractor (50 kW, 67 HP)
Weight, kg	55	750–1000	2500–3000
Pull, kgf	–	100	1500–2000
Speed, m/s (m/h)	–	1 (3.6)	1.7 (6.1)
Power, kW (HP)	0.07 (0.09)	1 (1.3)	25–34 (35–46)
Power requirements for mould-board ploughing, kg/cm^2		———— 0.7 ————	
Work capacity: implement size, cm, depth × width		10 × 14	20 × 100
work rate,[a] ha/h		0.04	0.45
work rate, h/ha	75–125	25	2.2
work day length, h	5	5–6	8–16
daily output, ha/day	0.07–0.04	0.20–0.24	3.6–7.2

[a] Assuming 70 per cent field efficiency.

a reasonable depth. In the time available, a pair of oxen could work about 3–4 hectares. Tractors can pull approximately half their weight. A 3000 kg tractor with a pull of 1500 kg force can manage a three-furrow plough. This means that a 50 kW tractor with about 25 kW at the drawbar could handle about 80 hectares per season. This aspect of Newtonian physics largely explains why small lightweight tractors, confounded by wheelslip, are often technically incapable of performing rudimentary cultivation work to an adequate standard.

There are qualitative aspects to the power criteria. Maximum power ratings can only be achieved for about two hours and five hours for men and animals respectively, whereas with proper maintenance engines will keep turning. Engine power is potentially also more adaptable and manageable.

In situations where energy is particularly limited, traditional smallholder systems are seen to be relatively efficient. The small farmer consumes about 0.1 kW/ha of mainly renewable energy; the UK farmer about 1.7 kW/ha of mainly non-renewable energy. The energy conscious will point to an energy-out to energy-in ratio of 10.5 for cereals on the Nigerian smallholding, and 2.4 for the UK farmer (Leach, 1976). Food production, however, varies in the opposite direction, at 0.8 tonne and 6 tonnes per hectare respectively.

8.6.2 Work rate and capacity

Peak power requirements are more important than averages. The size and nature of the task, and the time available, will together determine peak power requirements. Where time is limited, the rate of work of alternative systems of mechanization, as determined by their power characteristics, will be an important factor in selecting mechanization.

Seedbed preparation is often a critical period. Depending on local conditions, manual labour can do the job in fifteen to twenty-five days per hectare (Table 8.2). An ox team will take four to five days and a tractor about two hours of field time. Assuming a ten-to-fifteen-day planting season these capacities accord with those based on power requirements. Where timeliness penalties are high, as for instance during the cross-over period between irrigation seasons, the provision of high-powered, high-output, albeit expensive forms of mechanization could be justified. However, experience shows that the actual performance of tractor hire services, for instance, is much below theoretical capacity, due to management shortcomings and an environment that is often hostile to tractor use. Reducing the peak on one task may serve no more than to transfer it to another. Weeding particularly is an operation that is often overlooked in part because it is more difficult to mechanize or schedule.

As previously mentioned, there are opportunities for improving the productivity of existing power sources rather than switching to higher order ones. Introduction or modification of hand tools, such as rice planters and scythes, or the use of more advanced devices such as seeders, knapsack and ULV sprayers are cases in point. The development of ox tool bars and attachments provides other examples (ITDG).

8.6.3 Costs

Making a cost comparison of alternative mechanization systems is difficult because first there may be qualitative differences in the task performed and secondly the definition of costs will vary according to the purpose of the analysis. On the latter issue, a basic distinction needs to be made between financial and economic costs. The mechanizing farmer will be concerned with financial prices as these are what he pays for equipment or services in the market place. The government, however, will be more interested in economic prices which reflect the scarcity value, before taxes and subsidies, of the mechanization inputs to the economy. A rational mechanization policy would seek to bring the two price bases together.

A cost analysis of land preparation costs for smallholders in northern Nigeria, using government tractor hire services, hired or owned-oxen, and manual labour is contained in Table 8.3. In financial terms, labour appears least expensive where there is family labour with nothing else to do. But hired labour can be expensive and scarce during busy periods. Here, as in many other places, government tractor services are heavily (80 per cent) subsidized. Private contractors charge twice the price but still do not recover full costs. In general tractor services are limited in availability, and mainly directed at the larger farmers. The costs of owned-oxen vary, mainly with the level of use and the cost of feeding. Actual oxen hire rates depend on whether farmers reciprocate other favours. Full ox hire rates are often high, in this case reflecting a small work-bull population and the high cost of maintenance. Full hire costs and the full cost of properly fed and equipped owned-oxen are usually very similar.

Viewed in economic terms, adjusting for subsidies, taxes, shadow values of labour, fodder crops, and foreign exchange, tractors appear the most expensive and labour the cheapest method of work. During busy periods when labour is

TABLE 8.3 Comparative Land Preparation Costs, Northern
Nigeria
N/ha (1978 prices)[a]

	Financial cost to the farmer	Economic cost[b]
Tractor	10[c] (45)[d]	58
Oxen	20–39[e]	20–29
Hand	0–30[f]	15–30

[a] NAIRA (N) = £0.8
[b] Adjusted to allow for taxes, subsidies, shadow prices of foreign exchange, fodder and labour.
[c] Subsidized hire rate.
[d] Full financial cost including administration.
[e] Hire or ownership costs, depending on feed costs.
[f] Depending whether unpaid family labour or hired labour.

scarce, particularly on larger farms, the use of oxen power may become relatively attractive.

This order of ranking is common to most smallholder situations, although animal power tends to be less expensive where local conditions (particularly fodder availability) favour livestock. In Egypt for instance, with a heritage of working animals, land preparation costs in 1980 were LE 8.8 (about one-third of full costs) per hectare by the co-operative tractor, LE 12.5 by the private contractor, about LE 25 by oxen, and about LE 20 for hired labour ($£1 = LE$ 1.50). Private contractors were not fully recovering costs, and the condition of their tractors reflected this. Understandably, the demand for tractor services exceeds supply.

This simplistic analysis can be misleading. For instance, costs per unit of work vary according to how intensively the animals or tractor are used throughout the year. Government tractors (noted for broken hour-clocks) often achieve only 500 hours per year. Private sector tractors often work more than 1000 hours per year. Furthermore, there may be differential benefits between systems arising from quicker or more effective operations. With respect to the latter, however, in the context of mechanized cultivation, there is little evidence to show that tractor cultivation in itself provides long-term yield improvements. Where power is the constraint, and cost per usable kW is the criteria, there is little to beat the conventional tractor. Attempts to develop cost effective small tractors bear witness to this (Morris and Pollard, 1981).

Even where more mechanization is deemed cost-effective and worthwhile for the private farmer, it may not be within his financing, cash flow capability. Mechanization packages, on cash or credit, would need to be designed with this in mind.

8.6.4 Institutional support

Mechanization, like any other input, requires institutional support in the form of input supply, credit, extension and training, and adaptive research. Generally, the more capital-intensive and the less indigenous the technology, the greater are the demands for support services. Improved hand-powered systems are most easily supported. Establishing oxen programmes requires the provision of oxen, implements, medium-term credit, veterinary, and extension and training services (Mettrick, 1977). Tractors, whether from the public or the private sector, require the highest order of support services, particularly with respect to spare parts supply (Dalton, 1976). Existing advisory services are often biased towards agronomists and livestock husbandry specialists. An effective mechanization extension service requires conscious efforts to improve the provision of agricultural engineering specialists at technical and professional levels. Given the complete absence or low level of institutional support in many smallholder situations, the most appropriate mechanization system is that which is most reliable and self-sufficient.

8.6.5 The social and economic impact of mechanization

The effect of mechanization on labour employment, and related issues such as rural income levels and distribution, is a hotly debated topic, supported by a weighty literature. Improving hand or expanding animal systems is generally considered to have a gentle and mainly beneficial effect on employment and other socio-economic parameters (Eicher, Zalla, Kocher and Winch, 1970; Yudelman, Butler and Banerji, 1971; Mettrick, Roy, and Thornton, 1976; Bartsch, 1977).

The available evidence suggests that where mechanization facilitates an expansion in the cultivated area, an increase in cropping intensities, new crop mixtures, and the use of other improved inputs, overall employment can be increased. In many smallholder situations operating at low power input levels this is often the case (Roy and Blase, 1978; Bala and Hussain, 1978; El Hag, 1975; Giles, 1975; Clayton, 1973; Yudelman, Butler, and Banerji, 1971). After a certain point, however, further mechanization begins to substitute for labour, particularly permanent hired labour. The big debate is primarily concerned with defining this point. Extensive tractorization is usually associated with labour (and animal) displacement. Estimates for Latin America suggest that one tractor has replaced about five horses and twenty men (Abercrombie, 1973). In parts of Pakistan, a tractor has been associated with the net loss of eight permanent jobs (McInerney and Donaldson, 1975). The effect on labour employment has varied between types of worker; family and permanent hired labour inputs have tended to decrease whilst in some cases this has been partially offset by increases in casual labour requirements (Kahlon, 1975; Singh and Gaswami, 1977). Some researchers, observing the impact of technology on women, argue that, far from their being beneficiaries, the role of women can deteriorate as a result of farm mechanization. Men are often quicker to associate with machinery, leaving the unmechanized and most tedious jobs to women.

Tractorization may also have implications for agrarian structure, particularly farm size and tenure systems. As the level and sophistication of mechanization increases the smallest farmers find they are not big enough to adopt the new output-increasing (and possibly cost-saving) technology. In some cases they may be squeezed out by falling product prices due to the increased output generated on the larger, more successful farms. Unviable small farms are eventually amalgamated into larger holdings and their occupants become landless labourers. In some instances, landlords may dispossess their tenants in order to achieve economies of scale in machinery operation (Yudelman, Butler, and Banerji, 1971; McInerney and Donaldson, 1975).

Where tractorization projects have been subject to cost-benefit analysis, it is suggested that mechanization is economically profitable to the nation as a whole, and generally much more financially profitable to participant mechanizing farmers (Era, 1979; Dalton, 1976; McInerney and Donaldson, 1975; Gotsch, 1975). Many studies, however, vacillate between whether labour savings should be measured as an economic cost or a benefit, and the outcome of the analyses can be particularly sensitive in this assumption. The long-term social and economic implications depend on whether displaced farm labour can find gainful employment elsewhere. In many Third World economies this may not be possible.

8.7 THE BEST SYSTEM: MECHANIZATION POLICY

Which is the best course of farm mechanization to follow depends on the objectives of agricultural development, and the prevailing resource/constraint environment. Agricultural development strategies generally show a commitment to increasing production, rural incomes, and employment opportunities. Given the nature of most developing economies and their smallholder sectors, the most appropriate system of mechanization would be that based on a readily available, renewable and self-sufficient, and therefore low-cost, power source. Unfortunately, price structures in many developing economies are often imperfect and do not necessarily lead to the best use of available resources. And farm power is no exception. For example, overvalued currencies, government subsidies to tractor manufacturers and users, duty-free concessions to importers, cheap credit (particularly during times of inflation) and tax allowances on operating costs, may encourage the adoption of imported capital-intensive mechanization projects, which essentially substitute scarce and expensive capital for a plentiful and otherwise unemployed labour force. Labour, even where it attracts a barely subsistence wage rate, is often overpriced in real economic terms, which may further encourage wanton mechanization.

Given the role of mechanization in the process of getting agriculture moving, and its important social ramifications, mechanization policy becomes an important aspect of agricultural planning. In an attempt to remove smallholder power constraints, while avoiding the wasteful and undesirable effects of overmechanization, particularly labour displacement, many governments have embarked on a programme of 'selective mechanization'. This involves, by means of detailed farm management and agricultural engineering study, the identification of power peaks and the most appropriate, cost-effective way of dealing with them. Whilst some cynics regard selective mechanization as a conspiracy by engineers to do no more than placate welfare economists, proponents see that it provides a basis for a mechanization strategy involving a technical, financial, and economic assessment of the feasibility and the justification for mechanizing particular operations in given farming systems (Stout and Downing, 1975; Voss, 1975).

For a given farming system, selective mechanization would attempt to exploit the potential benefits of mechanization as previously enumerated; opportunities for increasing the area cultivated, the timeliness of operations, cropping intensity (multicropping), the quality of work, and labour employment. Consequently a selective mechanization approach may incorporate all three technology types; for example oxen for land preparation, manual harvesting, and engine power for water-lifting or thrashing.

In an attempt to achieve a rational approach to mechanization, a number of developing countries have devised smallholder mechanization strategies within the context of an overall development programme. For example, an ongoing US$40 m mechanization project in Egypt (USAID, 1979) has the following components:

(1) Research and development into farm power needs; inventory of existing machinery, adaption, improvement, development and testing.
(2) Mechanization planning and extension management.
(3) The provision of selected mechanization services (manufacture, maintenance and repair, operations) through public sector or government-assisted private sector operators.
(4) Formulation of selected smallholder mechanization 'packages'.
(5) Institutional support measures; pricing policies (taxes and subsidies), supply, credit, assistance to machine sharers, training and extension, demonstration, workshops.

Such a comprehensive and problem-oriented approach to mechanization technology choice and policy-making remains the exception rather than the rule. There are signs, however, of an improved rapport between engineers and economists, and a realization on behalf of policy-makers of the effect of mechanization on the nature and rate of agricultural and rural development, such that mechanization components will become a more important feature of future development projects.

8.8 CONCLUSION

At present levels of productivity, it is debatable whether the average smallholder should give priority either to yield-increasing inputs such as improved seeds and fertilizers, or to mechanization. In practice the two are often inseparable. The use of improved inputs provides the potential and justification for more farm power. Simultaneously, more farm power may be necessary before the potential of new yield-improving inputs can be realized. It is difficult to give a general answer to the question 'Man, Animal or Engine?' In some situations, all three power sources may be appropriate. From a policy point of view this requires a careful assessment of mechanization needs, an appraisal of available technology, and the formulation of policy measures which would encourage the development and selection of mechanization appropriate to the pre-defined development objectives.

REFERENCES

Abercrombie, K. C. (1973). Agricultural mechanisation and employment in Latin America, in *Mechanisation and Employment in Agriculture* (ILO, Geneva).
Bala, B. K., and Hussain, A. (1978). Farm size, labour employment and farm mechanisation in Bangladesh, *Agric. Mech. in Asia*, **11** (3).
Bartsch, W. H. (1977). Employment and technology choice in Asian agriculture (Praeger, New York).

Clayton, E. S. (1973). Mechanisation and employment in East African agriculture, in *Mechanisation and Employment in Agriculture* (ILO, Geneva).

Dalton, G. E. (1976). British Aid tractors in India; an *ex-post* evaluation, ODA (UK).

Eicher, C., Zalla, T., Kocher, J., Winch, F., (1970). Employment generation in African agriculture, Institute of International Agriculture, Michigan State University.

El Hag, H. E. (1975). Mechanisation production and employment in the Sudan. Report of the Expert Panel on the Effect of Mechanisation on production and employment in agriculture, FAO, Rome, 4–7 February.

Era 2000 Inc. (1979). *Further mechanisation of Egyptian Agriculture* (Gaithersbury, Maryland, USA).

Giles, G. W. (1975). The reorientation of agricultural mechanisation for the developing countries, policies and attitudes for action programmes. Report of Expert Panel, FAO, Rome, 4–7 February.

Gotsch, C. H. (1975). Tractor mechanisation and rural development in Pakistan, in *Mechanisation and Employment in Agriculture* (ILO, Geneva).

ITDE. Agricultural equipment and tools for farmers designed for local construction: design notes. Intermediate Technology Development Group Ltd., Shinfield, Reading.

Kahlon, A. S. (1975). Impact of mechanisation on Punjab agriculture. Dept. Econ., Punjab Agric. University, mimeo.

Leach, G. (1976). *Energy and Food Production* (IPC Science and Technology Press).

Lonnemark, H. (1967). *Multifarm Use of Agricultural Machinery* (FAO, Rome).

McInerney, J. P., and Donaldson, G. F. (1975). The consequences of farm tractors in Pakistan. Working Paper 210, IBRD, Washington.

Mettrick, H. (1977). Oxenisation in the Gambia: an evaluation, ODM (UK).

Mettrick, H., Roy, S., and Thornton, D. S. (1976). Agricultural mechanisation in Southern Asia. A Report to ODA (UK), Reading University.

Morris, J. and Pollard, S. (1981). The place of small tractors in the mechanisation process. International Agricultural Development, November 1981.

Pollard, S. and Morris, J. (1978). Economic aspects of the introduction of small tractors in developing countries, *The Agricultural Engineer*, **33** (2), 29–31.

Roy, S. and Blase, M. G. (1978). Farm mechanisation productivity and labour employment: a case study of Indian Punjab, *Journal of Development Studies*, **14** (2).

Singh, R. P. and Gaswami, H. S. (1977). A comparative study of tractorised and bullock operated farms in Purnea District, India. Agro-Econ. Res. Centre, University of Allahabad, mimeo.

Stout, B. A. and Downing, C. N. (1975). Agricultural mechanisation policy. Report of Expert Panel, FAO, Rome, 4–7 February.

USAID (1979). Agricultural Mechanisation Project, 263–0031, Egypt. Preparation Report, Cairo.

Voss, C. (1975). Different forms and levels of farm mechanisation and their effect on production and employment. Report of Expert Panel, FAO, Rome, 4–7 February.

Yudelmen, M., Butler, G., and Banerji, R. (1971). Technological change in agriculture and employment in developing countries (OECD, Paris).

9 *Performance of Powered Field Equipment*

An early prototype
of a small tractor
designed for
developing countries

The basic function of a powered field machine is to move an implement in a controlled fashion. It is the implement which performs the agricultural task, whether it is a soil-engaging task such as ploughing or a non-engaging operation like spraying. The machine therefore has, at a minimum, to carry itself plus some equipment across the field in a way which satisfactorily retains soil and crop characteristics. At the maximum, the machine is required to exert considerable drawbar pull in order that a soil-engaging implement may perform the required disturbance work on the soil. Within these functions there exists a very wide range of possible machine designs. The major factors involved are discussed below.

9.1 LAYOUT

This factor describes the way in which the various important components of a machine are arranged in relation to one another. The 'power' components consist of the power source and the means of transmitting that power to the implement.

This usually involves engine, clutch, gearbox, drive shafts, belts or chains, driven axle, wheels or tracks. The 'control' components allow the implement to be positioned in relation to the soil or crop. These components include the steering mechanism, and the implement depth and lift control. Where manual control is implied (as it nearly always is), the operator can be thought of as part of the control system.

9.2 POWER

There are two important values for power concerned with a field machine. The first is the power *required* in order for the implement to function satisfactorily. This power may either be in the form of 'linear' (i.e. drawbar) power, defined as force multiplied by speed, or 'rotary' power, expressed as torque multiplied by rotary speed. In both cases it is convenient to express power in kilowatts (kW)

Then drawbar power(kW) = drawbar force (kN) × speed (m/s)

and rotary power(kW) = torque (kNm) × speed (radians/s)

$$\text{or} = \text{torque} \times \text{speed (in revs/min)} \times \frac{2\pi}{60}$$

The second important power figure is the amount of power *available in practice* from the power source (engine). This is the nominal engine power less any losses due to engine condition, less losses due to environmental factors (air temperature and altitude), less losses in the transmission system on the way to the wheels or power-take-off shaft, less losses (in the case of wheels) due to wheelslip and rolling resistance. These losses can, in combination, be very substantial and as much as 40 per cent of the nominal engine power may be lost in this way in quite normal field conditions.

The significant 'power' aspect of a field machine is, therefore, whether the power available is equal to the power required. If the available power is less, then the machine will not perform the required operation (at least, not at the desired rate). If it is greater, then the machine has overcapacity of power which, if substantial, can lead to inefficient operation.

9.3 SPEED

This is taken as speed over the ground and, combined with the effective width of the implement operation, is important in determining the rate of work of the machine (ha/h). In addition it is an important part of the power equation described above, since an adjustment of speed will usually assist in obtaining a closer match between the power required and the power available.

9.4 WEIGHT

In the case of traction it is necessary to apply a certain force on to the driven wheel(s) or tracks. This force can come from direct weight (including ballast and implement weight) or from transferred weight (arising from the moments set up when a drawbar pull is produced). Sufficient weight on the steered wheels is also required. Weight is also, however, a significant factor in increasing rolling resistance and sinkage, and in causing soil damage.

9.5 WHEEL SIZE

The other important factor in traction, wheel diameter, should ideally be as large as possible to increase tyre thrust and reduce rolling resistance. Larger tyres, however, are more costly and require a great ratio reduction within the transmission system in order to provide the desired low speeds for field operation.

9.6 IMPLEMENT MOUNTING AND CONTROL

The position of the implement in relation to the operator has an effect on his visibility and the ease of positional control which can be achieved. In fact, the conventional implement position behind the driver is almost the worst that could be provided from this aspect. An ideal position is ahead of and somewhat below the operator so that he is able to steer the machine and control the implement while looking ahead.

9.7 COST

The initial cost of a machine is very important, particularly in the smallholder situation where incomes and returns are not generally high. Over its life, however, a machine will incur costs due to depreciation, spares and repairs, fuel and lubricants, wages, insurance, etc. Often expressed as a cost per hour, it is more useful to think in terms of a cost per unit area of land operated on. In this way a small, cheap machine with short life and high fuel consumption and with a low rate of work may well incur a higher cost per unit area than a larger, more capable machine. The criterion is usually whether the area available is sufficient to justify the larger machine, and whether the purchase price can be met. For this reason a cheap, small, economical and *effective* machine is desirable, but is not easily provided in practice. One reason for this is that difficult operating conditions require good performance and robust design, and it is not easy to produce this cheaply.

9.8 DIMENSIONS

There are situations where compact overall size is important, in achieving field access for example or in working between obstacles such as trees, stumps, or rocks. Manoeuvrability also tends to be better with compact size, but ground clearance over crops and obstacles should ideally be good. Stability tends to increase with width. There are thus a number of conflicting factors involved in determining the overall dimensions of a machine for field use.

9.9 EXAMPLES OF MACHINERY AND EQUIPMENT

Although there are dozens of possible variations in layout between number of wheels, steering mode, operator location and implement type and position, machines tend to fall into one of a small number of categories. Examples are given, in each category, of small tractors past and present—mainly past, which illustrate the difficulty of designing effective machines of this type.

9.9.1 'Conventional' four-wheel ride-on machines (Figure 9.1)

Examples are Power King (USA), Speedex (USA), Agrale (Brazil), International Cub (USA), Buffalo (USA), Agro-Util (USA) and most Japanese small tractors (usually four-wheel drive).

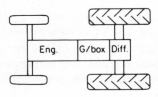

FIGURE 9.1 Schematic plan view of a 'conventional' four-wheel tractor layout

Based on the traditional four-wheel tractor, these machines often follow large tractor practice in having a T-shaped form (viewed from above) with cast engine/transmission casings forming the stem of the T and the cast rear axle housings forming the cross-member. Many medium and small two- or four-wheel drive tractors are of this type. Large-scale production is required and usually the conventional ancillary systems of large tractors are provided: electrical system and lights, hydraulic implement lift and controls, power take-off shaft and multiratio gearbox. Articulated four-wheel drive tractors also come in this category. Examples are Gordini (Italy), Agria (W. Germany), and Cast (Italy).

All the reliability problems associated with complex systems then occur, and in smallholder conditions the initial cost and operating costs may be very high. Because of the scaling down in size the performance of these machines is not likely to be adequate in difficult conditions.

There are some examples of medium-sized 'simplified' conventional type tractors being produced, particularly on the Indian subcontinent. These may have no electrics, are robust, simple and relatively cheap and, because of their reasonable size, will tend to have a fair performance. Examples are Eicher (India), and IRRI-PAK (Pakistan).

The third major category of four-wheel tractors is those that appear conventional (for reasons of acceptability) but are constructed in a very different way from normal. A fabricated steel chassis is usually provided, on to which various mass-produced parts such as the engine, gearbox, and axles are bolted. This has advantages in that the resulting machines may be partly manufactured and assembled locally, which can assist in reducing foreign exchange and parts availability problems. Specification changes to major components such as engines can be made fairly easily, to suit local conditions and component availability. Examples are Self-help (USA), Kabanyolo (Uganda), Howard (Australia), Winget (UK), Colt (UK), Croftmaster (UK) and Buffalo (UK).

Most of these machines, however, are fairly small and light, so that their performance can often be marginal. Ground clearance is often a problem, and the use of automotive rear axles may mean that no differential lock is provided, which does not assist the traction characteristics.

9.9.2 Unconventional four-wheel, ride-on tractors

A few manufacturers and design teams have accepted the deficiencies often involved in attempting to scale down large tractors, and have produced machines which do not particularly resemble them. Almost inevitably based on a fabricated chassis, these machines tend to have rear- or mid-mounted engines, and good ground clearance provided by final reduction drives (or in one case, by the use of hydrostatic transmission). See Figure 9.2.

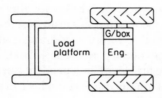

FIGURE 9.2 Schematic plan view of an 'unconventional' four-wheel tractor layout

A load platform is typically provided at the front of the machine, allowing a useful transport ability. Usually slightly larger and heavier than the other 'scaled down' tractors, their performance can be reasonable and, with good ground clearance and stability, provides rather better characteristics. There is no attempt to emulate the appearance of a 'normal' tractor, which may be a disadvantage with some potential customers. Examples are Tinkabi (Swaziland), Bouyer (France), and Russell 3D (UK).

9.9.3 Three-wheel machines

There are two major variants within the three-wheel class. The first has a single front wheel (Figure 9.3); the second has the single wheel at the rear, Figure 9.4. Both will be somewhat (but not necessarily significantly) less stable than four-

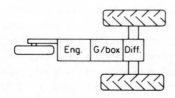

FIGURE 9.3 Schematic plan view of a three-wheel tractor with single front wheel

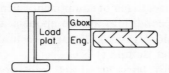

FIGURE 9.4 Schematic plan view of a three-wheel tractor with single rear (driven) wheel

wheel tractors. Both will have limitations on operation in row crops compared with a four-wheel machine because the track of the two wheels must span 2 rows at least (possible spacings are 2 rows, 4 rows, 6 rows, etc. compared with 1, 2, 3, 4, 5, etc. rows with a four-wheeler) in order to allow the single wheel to go down the middle. If, however, the crop row widths are established in the first place to fit the tractor track settings, this problem can be overcome.

The single front wheel type (US row crop) tractor has a potentially very simple steering system, but behaves in most other ways like a four-wheeler. It has two driven rear wheels and a differential. An example is Mouzon (France).

The single rear wheel type has very different characteristics. The absence of a differential simplifies the transmission design markedly but the steering system remains conventional. Two main problems occur in operation. The first is that good traction requires a large rear drive wheel with heavy loading and weight transfer. This, with a rear-mounted implement, can result in an unstable tractor. A front or mid-mounted toolbar is, however, a viable, low-draught alternative.

The second problem involves manoeuvrability at the headland. A 'conventional' four- or three-wheel tractor can be fitted with independent steering brakes which enable it to be swung round almost 'on the inside rear wheel' in a tight turn. A single drive-wheel tractor cannot operate in this way and can be difficult to manoeuvre in tight spaces. Examples are Poynter 'Triple' (Australia) and NIAE Monowheel (UK). Note also the side-car type Wageningen three-wheeler (Netherlands). The Bolens Ridemaster (USA) had a single driven *and* steered wheel at the front.

9.9.4 Two-wheel tractors

Otherwise known as single-axle tractors or power tillers, this mode implies a walk-behind type of operation (unless it is fitted with a trailer for transport or a ride-on toolbar for lighter draught operations, in which case it becomes a four-wheel, front-wheel drive, articulated machine).

There is, in passing, a second possibility, in the form of a two-wheel 'motor cycle' arrangement. A number of these machines are produced, primarily for rough terrain personal transport, and they usually have wide flotation type tyres. It is suggested by at least one manufacturer that such a machine could be used to pull an implement. The drawbar pull would probably be quite low, and the durability (the machines are usually based on lightweight motor cycle technology) would be suspect under heavy sustained power demands. However, this machine type is worth mentioning, because it could gain access to isolated fields by using only footpaths, could tow a trailer for transport operations and, in the field, could perhaps be fitted with a detachable wheeled outrigger to improve stability and control. Examples are Rokon (US) and Cob (USA).

Returning now to the single-axle type tractors, these may conveniently be divided into large and small. This apparently simplistic division is in practice realistic as these machines are usually either quite large, heavy and expensive (having 7–10 kW diesel engines, multiratio gear boxes and steering clutches), or quite small, light and cheap (having 3–5 kW petrol engines, two speeds and a solid axle)—Figure 9.5. These machines were discussed in more detail in Chapter 7. The large machines can be very tiring to operate, as a good deal of skill and effort is required from the operator.

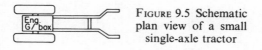

FIGURE 9.5 Schematic plan view of a small single-axle tractor

The small versions usually provide nominal traction and are best fitted with a rotary tiller blade (instead of or as well as the wheels), when their rate and quality of work can be reasonable in good conditions. Examples are Landmaster (UK), Geest (UK), Bouyer (France), Intec (USA), IRRI (Philippines), Krishi (India), Pasquali (Italy), and the Chinese single-axle tractor.

9.9.5 Winched equipment

The problems of limited traction, high slip and consequent heavy fuel consumption, often found with small-scale machines in high-draught operations, can be overcome by the use of the winch principle. A self-propelled winch moves across the field under the control of an operator. When the end of the field (or of the cable) is reached the drive to the winch is engaged, which causes the implement, attached to the end of the unreeled cable and controlled by a second operator, to move towards the winch. An anchor or 'sprag' on the winch unit prevents it from dragging, and the resulting operation, although fairly slow, produces a very high draught force at high efficiency from a light, low-powered machine (see Figure 9.6).

Extensive tests carried out in East Africa (Crossley and Kilgour, 1978) demonstrated the technical feasibility of this process, which enables an effective 'coefficient of traction' (i.e. pull divided by weight) of around 3.0 to be achieved with a small, two-wheel machine compared with about 0.65 with normal tractor operation. More recent tests with a larger, four-wheel ride-on version with the same engine power have demonstrated that the efficiency of winch operation can be combined with the versatility of a small tractor to provide a machine which can cultivate in hard soils at the rate of 0.2 ha/day with a fuel consumption of about 20 l/ha. It can also perform secondary operations and transport in the normal way by direct traction—see Figure 9.7. (Crossley and Kilgour, 1983.)

FIGURE 9.6 Small self-propelled winch system being tested in Central Africa

FIGURE 9.7 Prototype small tractor/self-propelled winch

9.10 PERFORMANCE

It is useful to be able to predict the performance of various machine types, in order to assess whether they are likely to be satisfactory in smallholder conditions. Chapter 11 gives detailed information on assessment and modification procedures. It is, however, possible to make reasonable predictions of some important performance factors as follows.

9.10.1 Pull

The drawbar pull produced by a machine depends on the characteristics of the machine and of the soil in which it is operating. A medium-sized conventional tractor operating at 20 per cent slip will pull about 75 per cent of its weight in good conditions. In very poor conditions the pull will reduce to about 40 per cent of its weight.

Smaller tractors are likely to have similar characteristics, but will tend to operate at higher slip values due to their smaller diameter tyres. This means that a tractor weighing 750 kgf will pull about 5.5 kN in good traction conditions and as little as 3 kN in very poor conditions. Since the required pull for a single implement can rise to 5 kN in hard, dry soil it is evident that many small, light tractors may not provide sufficient drawbar pull in these conditions.

9.10.2 Speed and rate of work

The speed at which a tractor operates depends either on the power available or on the speed selected by the operator according to noise and vibration restrictions. In high-draught operations it is likely to be engine power which limits the forward speed. In Section 9.2 it was pointed out that power = pull × speed. In the case of a pull of 5 kN being required at a speed of 1.5 m/s (5.4 km/h) the necessary drawbar power is 7.5 kW. With losses this can mean 12 kW at the engine.

A tractor with less than this engine power would have to operate more slowly. The theoretical rate of work (with an implement 300 mm wide) would be 0.3 × 1.5, that is 0.45 m²/s, or 1620 m²/hour, or 12,960 m² per 8-hour day. This assumes 100 per cent field efficiency, with no turn round or lost time. In theory 75 per cent can be assumed for field efficiency in typical conditions, giving a rate of work of 9720 m²/day or nearly 1 hectare per day. Practical results tend to be considerably lower than this, and 0.5 ha/day is often a reasonable estimate of work rate for a small tractor, with 2 ha/day for a larger one.

9.10.3 Fuel consumption

Medium-sized diesel engines use fuel at a rate of about 0.3 l/kWh. This means that a 50 kW tractor engine operating at full power will consume 15 litres of fuel per hour (although in many agricultural operations the engine will not be working at full power for much of the time). A small petrol engine may consume up to 0.5 l/kWh, and is more likely to be operated at full power. A 5 kW engine of this type would then consume 2.5 litres of fuel per hour.

As with cost per hour (Section 9.7), knowing the fuel consumed per hour is less useful than the figure for fuel per unit area of land covered. Because small tractors tend to operate at higher slip values than large ones, they will normally use a good deal more fuel per hectare (for example 35 to 70 l/ha, compared with 20–40 l/ha for a large tractor when ploughing). The winch-type machines mentioned in Section 9.9.5 when fitted with small diesel engines will consume fuel at about the same rate per hectare as a large tractor.

9.10.4 Stability

Smallholdings are more likely to exist on slopes than would be the case with large farms. Tractors and machines in many developing countries are not required to be fitted with roll frames or cabs, so that a certain danger to the operator exists in the event of sideways overturn. A supported object becomes unstable on a slope when a line through its centre of mass falls outside a line joining the 'downslope' supports.

In the case of a vehicle the wheels are usually (but not always) taken as the supports, so that a conventional 'rigid' four-wheel vehicle such as a motor car will become unstable (statically) when on such a slope that a vertical line through the centre of gravity passes outside a line along the ground joining the contact points of the downslope wheels (Figure 9.8).

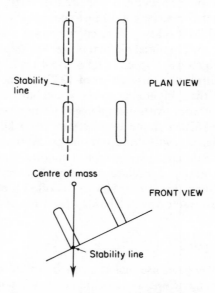

FIGURE 9.8 Stability of a 'rigid' four-wheel machine

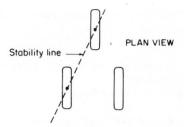

FIGURE 9.9 Stability of a three-
wheel machine

With a three-wheel machine the 'stability line' joins the contact point of the single wheel to that of the downslope member of the pair of wheels. Thus a three-wheeler is less stable than a rigid four-wheeler (Figure 9.9).

A conventional tractor, however, is in effect supported on its rear wheels and its front-axle pivot point. The initial stability then depends on a stability line 'in space' joining the pivot point B to the contact point A of the downslope rear wheel (Figure 9.10). If the pivot point is low, the initial instability (upslope rear wheel just leaving the ground) occurs on a slope not much steeper than that for an equivalent three-wheeler. When the tractor has tipped sufficiently for the front axle to come up against a solid chassis stop (if present) the machine will then behave like a rigid four-wheeler. In practice, however, it is often too late by then, as the tractor tends to carry on overturning due to the dynamic forces present.

If the pivot point is very high the front axle will overturn first. (This is never likely to happen in practice because pivot points are not designed as high as that.) It can be shown that optimum stability occurs when the axle pivot point is at the same height as the centre of gravity; the tractor then behaves throughout as a rigid four-wheeler. Detailed stability characteristics of various types of vehicles are as follows.

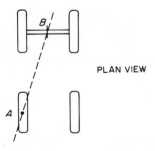

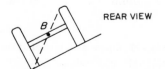

FIGURE 9.10 Stability of a
four-wheel tractor having a
pivoted front axle

TABLE 9.1 Vehicle Dimensions Required for Lateral Stability Programme
(metres)

	Symbol
Pivot point height (if pivoted)	$H9$
Horizontal distance of pivot point from rear axle	$H3$
Vehicle wheelbase	W
Height of centre of mass of vehicle plus payload	$H1$
Horizontal distance of centre of mass from rear axle	$C1$
Front track (where applicable)	$T8$
Rear track (where applicable)	$T9$

For three-wheel vehicles the limit of lateral stability is given by

$$\text{Tan}\,\beta = \frac{T(W - C1)}{2W\,H1}$$

where β is the slope angle, T is vehicle track (either $T8$ or $T9$). For four-wheel vehicles either with suspension (for example a conventional motor car) or with a pivot midway along the vehicle (for example an articulated tractor) the angle of limiting stability is given by

$$\text{Tan}\,\beta = \frac{T9W - T9C1 + T8C1}{2H1W}$$

Multiwheel vehicles such as all-terrain vehicles with six or eight wheels can normally be treated as four-wheel for lateral stability purposes.

For four-wheel vehicles with pivots near the front axle (for example a conventional tractor or a single-axle tractor with trailer) the limiting slope angle of the rear end is given by:

$$\text{Tan}\,\beta = \frac{T9\,(W - C1)}{2H3\,(W - C1\,H9/H1)}$$

while that of the front end (mass assumed small in comparison with the payload end) is given by $\text{Tan} = T8/2H9$.

9.10.5 Mobility

Mobility is taken as the ability to make progress in the conditions existing at the time. A bar to progress occurs when the machine is unable, for whatever reason, to maintain an acceptable speed across the ground. The main reasons why lack of progress can occur are as follows.

(i) Loss of traction. The thrust *able to be developed* by a wheeled tractive device depends, as stated previously, on a combination of wheel size and loading, and soil conditions.

The thrust *required* to move a machine in field conditions is the sum of any drawbar pull required (from an implement or trailer) plus the force required to overcome the rolling resistance of the tractor wheels themselves, both driven and undriven.

In difficult conditions and particularly where a machine is acting as a load carrier the rolling resistance alone may reach such a level (up to 30 per cent of the machine weight) that the available tyre thrust is not sufficient to be able to move the machine; in this case either the drive wheels spin or, if a low enough gear is not available, the engine stalls. This is complete lack of mobility.

(ii) Obstacles. Should the chassis or major components of a machine encounter an obstacle which itself has sufficient strength to produce a high resistance to motion, a similar situation will occur as in (i) above, and the machine will cease forward movement.

Obstacles can be divided into those which, in accumulation, provide resistance to motion (examples are standing crops or vegetation, acting against the underside or wheels of a machine to produce a resistance) and those which act substantially alone (such as stumps, boulders, fallen trees, ditches, and banks). The latter type of obstacle is usually divided into two classes causing, respectively, 'nose-in failure' and 'hangup failure'.

These descriptive terms refer to the action of a machine when encountering the obstacle. With 'nose-in failure' (NIF) the front (or rear) extremity of a machine grounds on a bank or bund over which it is attempting to climb and it is prevented from continuing, Figure 9.11. NIF does not normally occur with tractors because their wheels, both front and rear, are at the extremities of the machine, but it can certainly occur with vehicles. A tractor carrying a long implement which cannot be raised sufficiently high could suffer from rear NIF while traversing a ditch or bank.

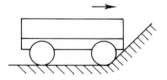

FIGURE 9.11 Nose-in failure of
a vehicle

'Hangup failure' (HUF) involves a part of the underside of the machine encountering an obstacle raised above ground level (such as a boulder, bund, stump, etc.). With vehicles or small tractors, HUF may occur under an axle or under the belly of the machine (Figure 9.12). Larger tractors are unlikely to suffer significantly from this failure in most conditions because of their good ground clearance, but due to their width they may have their progress interrupted by two separated obstacles such as stumps, between which a smaller machine could pass.

In the field with small machines the most likely mobility failure, apart from complete loss of traction, is hangup failure under the rear axle due to the relatively small rear wheels. Even 'dropped' final reductions to the wheels, designed to

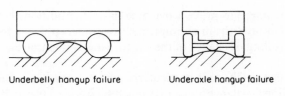

FIGURE 9.12 Hangup failure of a vehicle (two modes)

overcome this problem, may be prone to a form of hangup failure when, for example, a small tractor is mouldboard or disc ploughing with one wheel in the furrow. The bottom of the final reduction casing may then be very close to the ground, so that either a small obstruction or a momentary digging in of the drive wheel can cause hangup failure to occur. The basic problem with hangup failure is that, as the obstruction starts to support part of the weight, the traction tyre has less thrust and tends to spin and dig itself in (Figure 9.13). This increases the share of weight carried by the obstruction and in a matter of seconds the machine can be well and truly 'hung up'. (Note that detailed equations covering mobility may be found in Crossley, 1982).

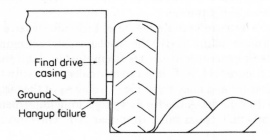

FIGURE 9.13 Hangup failure of a small tractor under final drive casings

9.10.6 Maintenance and life

The design life of a machine can be defined in engineering terms as the period during which the performance remains sufficiently close to the original 'new' performance to be acceptable. This definition is variable, depending as it does on the user's interpretation of what is acceptable. A machine with an excess of performance when new is more likely to remain acceptable than one which is on the limit of performance at the start.

An alternative (financial) definition can be taken as the period over which the machine has been 'written off' economically. It also implies that the operating costs are now too high to be acceptable. The major component of these costs with an older machine is likely to be the cost of maintenance, spares, and repairs. Given that a machine remains structurally sound it is, of course, possible to keep it operating almost indefinitely by replacing components with new ones as they fail.

Perhaps a realistic design life can be expressed as the period during which the major components of a machine (particularly the engine and transmission parts) will continue operating reasonably well with routine maintenance and overhauls. The importance of the latter factors cannot be overestimated. Without maintenance most machines will operate only for a relatively short period, reducing as conditions deteriorate. In very dusty conditions, for example, it is essential that an engine is provided with filters for the inlet air, and filters will choke in these conditions very quickly, causing severe performance loss. Should the choked filter rupture or be bypassed, the engine will become permanently damaged in a very short time.

The absence of skilled maintenance personnel, the high cost of spares and consumables such as oil and the difficult operating conditions existing in many developing countries combine to reduce machine life in some cases to a small fraction of 'normal'. Tractor 'lives' of two or three years are quoted in some cases. The depreciation then becomes a major factor in the total operating costs. For example, a tractor costing £9000 written off over three years has depreciated at the rate of £3000 per year. This is likely to be more than labour, fuel, and other costs combined.

Even where adequate maintenance is provided it may be found that the effective life of, for example, a small, high-speed petrol engine is between 500 and 1000 hours at full load. Where a machine is used for more than a few hundred hours a year this can mean a very short life, unless the engine is regarded (in the same way as items such as tyres) as a 'consumable' component which will be replaced during the overall life of the machine.

9.10.7 Effectiveness

Effectiveness implies performing the tasks required at a reasonable work rate and cost. Chapter 11 describes methods of assessing the performance of a given machine in relation to the task. Some general principles of effectiveness are given here.

(1) High-draught operations such as chisel ploughing, disc and mouldboard ploughing in dryland conditions may require a considerable drawbar pull even for a single implement. In good traction conditions a small tractor (weighing approximately 1 tonne and with an engine power of about 10 kW) will probably be satisfactory, although the rate of work will not be particularly high. Typical drive tyre sizes are 7.50–16.

Where traction conditions are poor (for example where the soil surface is wet or loose) a tractor of that size will probably not operate satisfactorily and a heavier machine (2 tonnes minimum) with larger drive tyres (10–28) would be required. A conventional medium–large tractor is likely to operate effectively in most soil conditions, provided that the field size and distribution is such that it can achieve a reasonable utilization efficiency.

Very small or 'garden'-type tractors are very unlikely to be heavy enough or to have large enough drive tyres to give good results in any but ideal conditions. Fuel consumption (probably with a petrol engine) would be high and durability poor.

(2) Low-draught operations such as weeding, spraying, and harvesting can probably be done reasonably well with most types of tractor, provided it has sufficient mobility to move within the existing field conditions. In practice this usually means that it should have wheels large enough (7.50–16 minimum) to traverse ploughed land, and a ground clearance sufficient (around 400 mm) not to damage standing crops.

(3) For transport operations the forward speed of some small tractors may be regarded as too low; 20 km/h is a reasonable speed in top gear. Large tractors normally do not have inbuilt transport facility, and a large two- or four-wheel trailer of 5–10 tonnes capacity makes an effective transport device. This is an expensive set up, however, and in the smallholder situation it may be found that a small tractor with trailer or load platform is a viable alternative.

Single-axle tractors fitted with trailers are effective transport devices provided traction conditions are reasonably good. They are not, for example, likely to perform well in the field when loaded, particularly if the soil is disturbed, because the load is carried on the undriven wheels.

REFERENCES

Crossley, C. P., and Kilgour, J. (1978). Field performance of a winch-powered cultivation device in Central Africa, *J. Agric. Engng. Res*, **23**, 285–396.

Crossley, C. P. (1982). Rural transport in developing countries—the development of the 'CARTA' computer program, *J. Agric. Engng. Res.*, **27**, 139–53.

Crossley, C. P., and Kilgour, J. (1983). The development and testing of a winch-based small tractor for developing countries. *J. Agric. Engng. Res*, **28** (2) 149–161.

10 *Operator Safety and Efficiency*

A truck has failed to negotiate a curve on a dirt road in the wet season

10.1 INTRODUCTION

There are a number of machines which are not continuously controlled by an operator. Examples are diesel or electric irrigation pumps, wind-driven devices, some processing tasks, large-scale irrigation equipment, etc. The characteristic of the tasks performed by this equipment is that they are continuous processes either with uniform performance demands or with only slight variations which can be anticipated and programmed for. Major variations cannot be accommodated.

Most field operations are clearly not of this type. Variations in conditions and constraints such as other crops, field boundaries, and weather considerations mean that while performing field operations the speed, location, and direction of the machinery must be continuously monitored and corrected by varying the settings of the controls. The decisions called for in this process are well beyond the ability of a complex computer (at least, at present technology levels). The problem is not so much dealing with the information, processing it and using it

201

for a control function—these tasks can easily be dealt with by a microprocessor; the problem is knowing what to look for, deciding what is important, assessing the magnitudes of the various data and then making it available for control purpose. It is just this kind of complex assessment, compilation, and decision-making which a human operator can handle so well. That is why, in the vast majority of cases, machinery required to perform 'dynamic' tasks such as field operations will be controlled by a human operator. Consequently it is important to be aware of the limitations which the machine may place on the performance of the operator, since most limitations will normally be fed back into the control of the machine, resulting in lower than optimum performance.

The topic of the operator as part of the machine system will be dealt with under the headings of safety and efficiency, although in practice there will be a number of instances where an overlap will occur.

10.2 OPERATOR SAFETY

The term safety will in this context be taken as meaning 'lack of damage' to the operator. Damage will be divided into two classes—immediate and long-term.

10.2.1 Immediate damage

This kind of damage will involve an event which takes place within a few seconds and produces immediate and (usually) obvious damage to an operator. In most cases it will involve contact with a part of the machine. Contact with the ground is not ruled out as a source of injury, but most agricultural machines travel slowly on fairly soft surfaces. Simply falling off a tractor is not likely to produce serious injury. The injury normally occurs when contact with part of the machine is involved during the process of leaving it. The classic example is where a tractor overturns on a slope. If the operator is thrown (or jumps) clear he is unlikely to be hurt. Unfortunately in many cases the tractor will fall or roll on to him in an overturn situation. This event was responsible for an average of thirty deaths per year in the UK alone until the fitting of roll cages or cabs became mandatory. These protective devices are seldom fitted to tractors used in developing countries, so the potential for injury or death is still high in these circumstances.

The second major cause of injury is caused by falling from a machine into an associated moving part. Falling from a tractor into the path of a trailer or towed implement has caused horrifying consequences (particularly to children riding as passengers) so the carrying of passengers on tractors is also banned in the UK. Again in developing countries the attraction of self-propelled machinery often causes additional passengers to be present, with consequent possibilities of injury.

The third situation that may arise is where the operator inadvertently touches a moving part of the machine such as a fan, belt, chain, shaft, or blade. Proper guarding can reduce the chance of this type of injury occurring, but guards can be removed and the final answer is likely to be a combination of control interlocks (for example preventing a tractor from being started unless it is in neutral and the clutch is disengaged) and driver education.

10.2.2 Long-term damage

This is taken as being damage which will not seriously affect a temporary operator but will eventually cause lasting damage to an operator who spends a number of years driving the machine. He will often be unaware of the damage taking place. The two major areas involved are the effects of noise and vibration.

(i) Noise. Noise damage can occur due to a sudden very loud noise or to sustained 'normal' loud noise. The first occurs at a sound pressure level ('noisiness') of over about 120 decibels (dB), and is very unlikely to result from agricultural machinery; it would, in any case, properly be categorized within Section 10.2.1 as immediate damage. Sustained 'normal' loud noise is (unfortunately) far more common. Because individuals vary it is not possible to define a reasonable 'absolute' level of noise below which no-one would ever experience hearing damage in the long term. The level currently taken is around 85 dB. At this level about 95 per cent of the population exposed to it continuously would not suffer damage. When it is realized that most tractors without cabs will produce noise levels at the operator's ear in the range 90–100 dB (Matthews, 1971) and that small high-speed engines fitted to small tractors can produce levels of over 100 dB (Crossley, 1979) it can be seen that a considerable amount of long-term hearing loss can occur with operators of agricultural machinery.

It can be argued that most machinery is not operated continuously day in and day out. Unfortunately the 'cancelling out' effect of periods of low noise is not very significant. Periods of loud and quiet noise can be summed to give an equivalent continuous sound level (ECSL), that is the continuous noise which would produce the same damage. For example a level of 100 dB experienced for only 1 hour per week is equivalent to a continuous noise of 84 dB; for 10 hours a week it is equivalent to 94 dB; and for 20 hours a week the equivalent level is 97 dB. (This is assuming total peace and quiet in the intervening periods.)

It is worth outlining here the steps that could be taken to reduce the noise levels at the operator's ear. Fortunately in most field operations the effect on the 'bystander' may be ignored (since the creatures of the countryside are unlikely to stay around long enough to be deafened). Protecting the operator can be achieved by a series of steps from the source to the operator. In sequence these steps are:

(1) Reduce the noise from the source—not an easy process without expensive redesign.
(2) Contain or isolate the source, for example by placing the engine in an enclosure. The key features concerning the enclosure are *denseness* of material and *completeness* of the enclosure. Because sound travels around corners (unlike light) it will escape through a gap and attack the operator even if he is 'shadowed' by a wall.

(3) Reduce the transmission of noise along frames, conduits, drive lines, etc. This can most easily be achieved by mounting the source and/or the operator's workplace on resilient mountings such as rubber.

(4) Reduce the amount of airborne noise reaching the operator by enclosing him, for example in a cab. It should be noted that a glass and metal cab of the weather-protective type will *not* normally reduce the sound level. It is more likely to increase it, by up to 10 dB (Matthews, 1973) due to reverberation of sound within the cab. Again the enclosure should be complete and should also provide for the *absorption* of noise entering the cab, through the use of foam linings and perforated coverings. The modern tractor 'Q' (for quiet) cabs provide lessons in this area.

(5) Reduce the noise *entering* the operator's ears by using ear plugs or muffs or both. These work better at high frequency than at low frequency, which is fortunate because the maximum sensitivity of the human ear is at quite a high frequency, around 4800 cycles/second (Hz).

(ii) Vibration. Vibration inputs above a certain level when transmitted to a machine operator can cause long-term damage. Vibration damage occurs most at around 4 Hz, which is the frequency at which the internal organs tend to resonate within the chest cavity. Damage to the spine also appears to result from exposure to high levels of vibration at this frequency. Unfortunately the natural frequency of tractor tyres is also usually around 4 Hz and since tractors are not fitted with suspension the resultant vibration transmitted to the operator during operation on most agricultural surfaces will be likely to produce damage in the long term. Recommendations for exposure limits depend on frequency and acceleration (magnitude) of the vibration. It is found that several agricultural operations (particularly secondary cultivation and transport) will exceed these levels significantly if performed over an 8-hour day (Figure 10.1).

It is not easy to eliminate harmful vibrations transmitted to a tractor driver, because the fitting of suspension to the machine itself is very difficult in view of the need to maintain a stable base relative to the soil, to which an implement can be attached. Front-wheel suspension is easy to fit but has relatively little beneficial effect since in most tractors the operator is positioned over the rear wheels. Suspended cabs have been investigated and represent a technically satisfactory solution, the economic cost of which is at present regarded as too high.

Suspension seats are the usual way of dealing with the vibration problem in agricultural machinery. A well-designed seat having a natural frequency of about 1.5 Hz and a damping factor of about 0.7 is likely to reduce vibration levels by 50 per cent (Matthews, 1973).

High-frequency vibration can occur with pedestrian-controlled machines having high-speed engines, or with other machines such as chain saws. Extended exposure to these levels of vibration can cause permanent loss of feeling in the fingers. Anti-vibration mountings and flexible hand grips can help to absorb this type of vibration input.

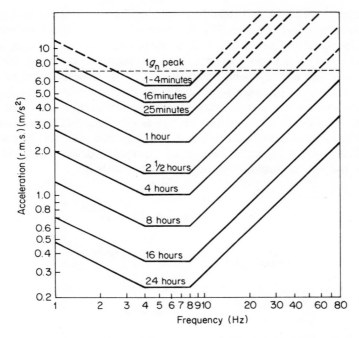

FIGURE 10.1 Exposure limits for vibration at various frequencies (Extracts from DD 32; 1974 are reproduced by permission of the British Standards Institution, 2 Park Street, London, W1A 2BS, from whom complete copies can be obtained)

10.3 OPERATOR EFFICIENCY

This term is taken to mean that the efficiency of machine control may be affected by constraints imposed upon the operator, usually arising from the design and operation of the machine itself. Some constraints, such as noise and vibration, may also cause damage as described in the previous section. When the operator is aware of the unpleasant effects of these inputs, he may take steps to reduce them by, for example, dropping the engine speed to reduce the noise and slowing the machine forward speed to levels where the perceived vibration is acceptable. There is considerable evidence (Matthews, 1982) that tractor operators tend to run their machines at lower than optimum efficiency in order to reduce the effects of noise and vibration. Other factors which tend to reduce efficiency of operation are mental and physical work load.

10.3.1 Mental workload

In order to maintain control of a machine an operator needs to receive input information using a variety of senses (sight, hearing, touch, smell), compare the resultant impression of the state of the machine with the desired state, decide whether corrective action is required and take such action in the correct sense so that any error is reduced to zero. This, in control terms, is called closed loop with negative feedback (Figure 10.2). Simple automatic systems such as a float-controlled valve on a water trough operate on the same principle. With field operations it has already been mentioned that the process requires a very sophisticated system in order to maintain accurate control. Fortunately the human operator when well trained can easily provide the required level of control but his efficiency does depend on whether he has good visibility, good instrument displays, and well-designed controls with which he can produce the corrective action. This overall aspect is known as workplace design. The subject is extensive, but we can summarize the important points here.

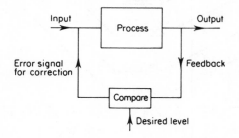

FIGURE 10.2 Closed loop control process
with negative feedback

Firstly, the operator should either be able to see what is going on, or he should have clear and easily read instruments to inform him. Important instruments should be mounted in front of and below his eye level. Secondly, the controls should be divided sensibly into hand and foot controls, and should be located in the areas shown in Figure 10.3. Less important or infrequently used controls may be located elsewhere (Matthews and Knight, 1971).

Thirdly, he should be comfortably positioned on a seat which is not only padded, but also ergonomically shaped as shown in Figure 10.4. The basic dimensions of the seat will vary depending on the task, but essentially the seat base (squab) should be inclined back at about 5° to the horizontal and the backrest at a further 100° (i.e. at 15° to the vertical).

10.3.2 Physical workload

A machine, particularly a power source such as an engine, has a certain designed rating. Provided that the machine is kept in good condition and given the necessary inputs (fuel and air in the case of an engine) it is likely to continue to operate at its rated output for 24 hours a day if necessary. In practice there are relatively few machines which do this, but irrigation pumps and expensive processing plant may be examples.

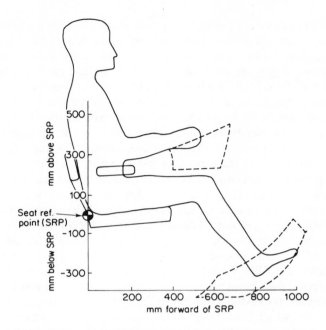

FIGURE 10.3 Recommended areas for important hand and
foot controls (after Matthews and Knight, 1971)

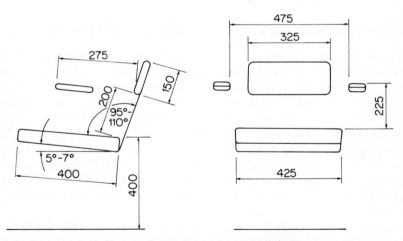

FIGURE 10.4 Recommended seat dimensions (after Matthews and Knight,
1971)

A human operator, on the other hand, is usually unable (or unwilling) to
operate consistently for more than 8–10 hours in the 24-hour day. This mismatch
can be overcome by providing say 3 × 8 hour shifts, and this is done in certain
situations. It is also important to realize, however, that the human operator or

worker also has a rated output and that, over an extended period of hours, this rated output is surprisingly low. It depends on the muscles obtaining a sufficient supply of oxygen, via the lungs and blood stream, to neutralize the results of muscular action. Failure to do so results in tiredness and pain, and is called the 'anaerobic limit'. This limit, in a healthy male, is typically an internal work level of 5 kcal/min and, translated into physical work output, results in a sustained power level of only 70 W. This effect can be obscured by the ability, shared by draught animals, to exert many more times this level (perhaps five times) for short periods. In the process, however, an oxygen debt is being built up, which will require repayment in the form of a rest period.

Therefore, if a work or control task requires a level equal to twice the anaerobic limit, the rest periods will have to total about the same as the periods of work. It can be seen that operations such as hand cultivation will inevitably be constrained by this anaerobic limit, and indeed the amount of land able to be cultivated by one person is limited for this reason (usually taken as 1/2 ha). It may also be found, however, that even the control of some large machines may require a high physical workload. If it is found that, due to high steering or pedal forces, the machine output reduces with time due to operator fatigue, it is worth considering incorporating servo assistance (or alternatively having a change of operators). Control of some small machines in difficult conditions, for example single-axle tractors in wetland rice cultivation, has been found to be limited to about 3 hours a day (Dibbits, Van Loon, and Curfs, 1978).

10.4 INCREASING OPERATOR SAFETY AND EFFICIENCY

Before the advent of farm machinery, traditional agriculture in developed countries consisted of a series of tasks carried out at various times of the year by hand or animal power. The limitation of damage to a human operator was the amount of work which he could perform without becoming excessively fatigued. There was, therefore, in most situations, a built-in restraint on operator damage. Similarly the types of accident which occurred were usually confined to falling off buildings or making contact with the wrong end of a pitchfork. With the increase in the use of machinery, as has been seen, the possibility of immediate or long-term injury has tended to increase, and it is only relatively recently that full attention has been paid to the reduction of damage and injury to agricultural workers.

As agricultural tasks in developing countries become more mechanized, there is the likelihood of a similar trend becoming apparent. In order to attract future generations to the land rather than to the urban centres it is necessary to reduce the drudgery and increase the interest of farming operators. Suitable mechanization can help to achieve this, but it is felt that the lessons on farm safety and operator welfare learnt so laboriously in the West may not be transferred effectively to the developing world unless attention is drawn to the potential problems.

10.4.1 Increasing operator and bystander safety

Tractor overturns can be prevented by correct operation which, in turn, can be encouraged by driver education. There are a number of reasons why tractors may be caused to overturn (the word 'caused' is used advisedly, since tractors do not turn over on their own). Taking corners at too high a speed, dropping a front wheel into a ditch while turning at the headland, and driving along a sideslope of too great an inclination are obvious causes of potential overturns. Work at the Scottish Institute of Agricultural Engineering (SIAE) has revealed another important feature of tractor behaviour, particularly with two-wheel-drive tractors (the normal type in developing countries). It concerns operation on grass-covered slopes or indeed on any slippery surface, where it is found that the most dangerous operation can be attempting to drive *down* a slope. The reason is that 'natural' weight transfer on to the front wheels of the tractor from the effect of the slope can reduce the grip of the rear (driven and braked) wheels to the point where they lock and start to slide. The best that can then happen is that the tractor may continue to slide straight down the slope and, if it flattens out, come to a safe halt. It is far more likely, however, that the tractor will start to turn as it slides and then a multiple overturn is almost inevitable. This event is completely un-controlled immediately the rear wheels start to slide and, because of the distance involved, can result in severe deformation even of approved safety frames. The important point for the driver to realize, therefore, is that a tractor can sometimes be driven quite safely up a slope and even along it, but the moment it is turned to go down the hill, complete loss of control may occur. The effect when a loaded, unbraked trailer is being towed is even worse.

Safety cabs are unlikely to be fitted to the majority of tractors exported to developing countries. However, simple and reasonably effective roll frames can be welded up from rectangular hollow section (tubular) steel and will serve also to support sunshades to reduce the effects of direct radiation. Provided that the operator is trained to hang on to the steering wheel and frame, in the event of a simple overturn he stands a reasonable chance of avoiding injury.

Tractors in developing countries are often used, during a considerable proportion of their lives, on transport tasks such as hauling trailers. Their good traction characteristics in poor conditions mean that tractors often provide the only viable choice for hauling trailers on estates and civil engineering projects. It must be remembered, however, that conventional tractors have no suspension and have brakes only on the rear wheels. Travelling at speed on poor surfaces can result in directional control which is no more than nominal, while braking in emergency conditions while hauling heavy (sometimes multiple) trailers, particu-larly on corners, may lead to severe consequences.

Some trailer hitches are designed to reduce the offset thrust during braking and, in the limit, to allow the trailer to overturn without taking the tractor with it. A preferable alternative will usually be four-wheel drive and, most important, four-wheel brakes.

Even when not used on haulage tasks, tractors are often driven on roads while journeying to scattered holdings. When carrying heavy, mounted implements the weight transfer effects may again reduce steered wheel loadings to below the desirable 10 per cent of total tractor weight for proper control, while the relatively slow speed and poor rear lights present a hazard to following vehicles. Reflectors are a useful supplement to rear lights (reflective plates are now obligatory on commercial vehicles in UK).

Finally, the practice of family labour groups on small farms means that all machinery is potentially dangerous, particularly to young people and children. Machine guards, education, and training are the major ways of reducing accidents. Legislation is unlikely to be effective in such conditions.

10.4.2 Increasing operator efficiency

When a new task is to be performed with a human operator it is tempting to carry out a short trial, pronounce the machine satisfactory, and plan on the basis of a similar long-term performance. Regrettably, this performance is not always forthcoming. Since the machine is not likely to be at fault, the effect of the efficiency of the operator's control is probably the culprit. It is pointless to insist that the operator improves his performance without trying to ascertain the reason for the shortcoming. Enquiries could be directed towards the area of the features he dislikes about the machine. This may enable problems to be discovered. For example, the cooling air outlet from an air-cooled engine may point towards the operator, as may a silencer, and ducting or repositioning will easily improve matters. High-frequency vibration can be reduced by the use of a resilient seat cushion (although low-frequency vibration cannot). High noise levels from a small engine can often be reduced to a more acceptable level by fitting a silencer from a bigger machine. (There may be a loss of efficiency, but this will occur anyway if the operator keeps the engine running at a lower speed in order to reduce the noise.)

Finally, it is the operator who is usually in a position to detect incipient faults on the machine, which uncorrected may lead to reduced efficiency or damage. A simple instruction to report all *changes* in the machine (noise, vibration, smell) can result in useful and early correction of faults, provided the spares/repairs facilities are available.

REFERENCES

Crossley, C. P. (1979). Theoretical design of small tractors, *Agricultural Engineer (UK)*, **33**, 2, 34–6.

Dibbits, H. J., Van Loon, J. H., and Curfs, H. P. F. (1978). The physical strain of operating a two-wheeled tractor, *World Crops*, November–December, 250–2.

Matthews, J. and Knight, A. A. (1971). Ergonomics in agricultural equipment design, NIAE, Silsoe.

Matthews, J. (1973). The measurement of tractor ride comfort, Soc. Auto. Engrs, Paper 730795.

Matthews, J. (1982). Effective choice and use of agricultural tractors, *The Agricultural Engineer (UK)*, **37**, 3, 96–9.

11 *Assessment and Modification of Equipment*

A winch-based prototype tractor being tested in East Africa—the disc shaped object on the lower chassis is a vibration type hourmeter.

11.1 MANUFACTURER-BASED TESTING

The principal reason for performing tests is to determine the performance of a device. This performance is in most cases required for technical comparison, either with the performance of an alternative device, or with performance prediction made at the design stage. Indirectly, however, the purpose of obtaining information through tests is to compare it with the requirement which the device was developed to satisfy.

It is all too easy to lose sight of this objective when developing a test procedure, and some manufacturer-based tests may be examples of this tendency. In the interest of accuracy and repeatability, a test may be evolved which, although fair, is not as relevant to the end user as it could be.

Perhaps the clearest example of this is in tests of tractor drawbar pull on concrete surfaces. Although some dried clay soils have been described as 'like concrete' it is not normally of interest to a farmer to know the traction capability of a tractor on such a surface. He is more interested in the performance characteristics on typical agricultural soil in various conditions. The problem is that agricultural soil is a very variable medium. It varies not only with type but also with time, moisture, temperature, and the sequence of previous operations. Performing comparable tractor pull tests on agricultural soils (a practice attempted from time to time by at least one farming magazine) is fraught with problems due mainly to variations in soil characteristics (angle of friction, cohesion, bulk density, moisture content, and surface characteristics).

The result has tended to lead to test procedures where tractors are tested on reproducible surfaces such as concrete, the conclusion being that even if the actual performance figures are not realistic for field conditons, they are indicative of such performance, so that farmers may compare different tractors before buying.

This is largely true with tractors of similar dimensions. A certain drive tyre size when loaded to a certain value will produce a definite pull at a given slip on concrete, regardless of which tractor it is fitted to. There have been examples, however, when a 'standard' test designed for large tractors has been applied to small tractors, the small drive tyre diameters of which then appear to be much less of a disadvantage, when tested on concrete, than would actually be the case in the field. The manufacturer is then at liberty to quote this unrealistically high tractive performance, which can mistakenly be accepted by potential buyers as actual likely performance. It is, in short, not easy to design a test which is both realistic and useful and which will give an immediate indication of the likely tractive performance of tractors in agricultural conditions.

Much useful work has, however, been done at NIAE, Silsoe, using a single-wheel tester to evaluate tractor tyres at various loadings in a range of soil conditions. These results are available in Dwyer, Evernden, and McAllister, 1976; an example of the smallest tyre performance (12.4/11-36) appears in Table 11.1.

TABLE 11.1 Tyre Performance at Various Loadings and Soil Conditions
(12.4/11-36 Tractor Driving Wheel Tyre

	Load:	kg	1180		1400		1600		1730	
		lb	2600		3090		3530		3810	
	Inflation	bar	0.8		1.1		1.5		1.7	
	pressure:	lb/in^2	12		16		22		25	
	Field conditions		kN	lb	kN	lb	kN	lb	kN	lb
Pull at	Dry grassland		8.7	1960	10.3	2320	11.8	2650	12.7	2860
20% slip	Dry stubble		6.1	1370	7.2	1620	8.2	1840	8.8	1980
	Wet stubble		5.8	1300	6.7	1510	7.6	1710	8.1	1820
	Dry loose soil		5.6	1260	6.5	1460	7.3	1640	7.7	1730
	Wet loose soil		4.8	1080	5.3	1190	5.7	1280	5.9	1330
Rolling	Dry grassland		0.9	200	1.1	250	1.3	290	1.4	310
resistance	Dry stubble		1.0	220	1.2	270	1.4	310	1.5	340
	Wet stubble		1.1	250	1.4	310	1.6	360	1.8	400
	Dry loose soil		1.2	270	1.5	340	1.8	400	2.0	450
	Wet loose soil		1.5	340	2.0	450	2.4	540	2.7	610

From Dwyer, Evernden and DcAllister, 1976.

11.2 USER-BASED TESTING

(Note: it is anticipated that the following sections will be relevant to industrial or quasi-government testing and research organizations, and not to a machinery user.) Agricultural machinery is becoming increasingly expensive. Even where a certain proportion of small-scale equipment can be made locally, the more complex components such as engines, bearings, belts, chains, gearboxes, etc. are likely to be imported, consuming scarce foreign exchange. Even local production uses materials, skills, and other resources which could be used for competing products. For these reasons it is becoming increasingly important to ensure that any equipment produced or imported in large numbers will be appropriate to the agricultural needs of the country. Since a large proportion of the agricultural production in many developing countries is of the smallholder type, the needs of this sector will be particularly important. As has been illustrated in earlier chapters, the conditions existing at the smallholder level are often not conducive to the easy operation of machinery in general, and in particular are far removed from those conditions in which much of the available large-scale equipment was designed to operate (Inns, 1978).

This means that it is not usually possible to extrapolate the performance of equipment from temperate agriculture to operation in the smallholder situation. Nevertheless, in order to avoid the inefficient use of foreign exchange and resources mentioned earlier it is important that the equipment selected will actually perform the tasks required in the conditions existing.

This can most effectively be established by evaluating examples of the equipment according to realistic test procedures. The establishment of such procedures is now discussed. (An overall summary of the procedure appears in Table 11.2)

TABLE 11.2 Evaluation Procedure For Machinery Assessment

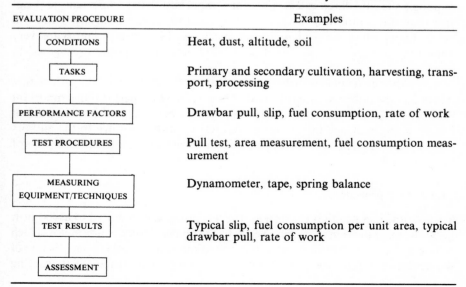

EVALUATION PROCEDURE	Examples
CONDITIONS	Heat, dust, altitude, soil
TASKS	Primary and secondary cultivation, harvesting, transport, processing
PERFORMANCE FACTORS	Drawbar pull, slip, fuel consumption, rate of work
TEST PROCEDURES	Pull test, area measurement, fuel consumption measurement
MEASURING EQUIPMENT/TECHNIQUES	Dynamometer, tape, spring balance
TEST RESULTS	Typical slip, fuel consumption per unit area, typical drawbar pull, rate of work
ASSESSMENT	

11.3 EVALUATION PROCEDURES

11.3.1 Tasks

The first stage in developing an evaluation procedure is to examine the *tasks* which must be performed and the conditions in which they will be carried out. For example, if the task is primary cultivation using a disc plough, to be performed during the dry season in smallholder fields of average size 0.5 ha, a good deal of information can be deduced about, for example, the kind of hazards to engine operation which will exist (dust, heat, altitude, vegetation), the drawbar pull required (upto 5 kN per disc), the amount of space available for turn round, the possible existence of ridges and other features, the time available for work, the quality of fuel which can be supplied, the levels of operator and maintenance skills and so on. Note that it is not only the technical aspects which must be considered, but also the social and economic infrastructure involved (Pollard and Morris, 1978).

11.3.2 Characteristics

The second stage is to determine the important performance *characteristics* required of whatever machine is to be considered. A limited time available for cultivation, for example, dictates a certain rate of work (ha/day) which in turn requires a satisfactory forward speed combined with implement width. This implement width in the particular soil conditions requires a specific pull. The performance requirement would then be a certain minimum drawbar pull at a given forward speed.

When a list of characteristics (say A, B, C, and D) has been drawn up it should be subjected to a 'weighting' procedure. The simplest way to do this is to divide the requirements into three categories—very important, fairly important, and not very important, allocating 'weights' of 5, 3, and 1 to each respectively. These will be used later when assessing individual machines. Let us say A and B are 'very important' (weighted 5), whereas C and D are not very important (weighted 1).

Note—although it appears rather academic to carry out this weighting procedure it is in fact the only way of ensuring that machines which perform the important functions well get more credit than those which perform well in unimportant functions. At least one (otherwise useful) test procedure known to the authors gives an equal weighting to all factors; this can severely reduce the effectiveness of the test results.

11.3.3 Test procedures

We are now in a position to consider the test *procedures* required to assess how good the machine is in performing each of the above factors. Some factors (such as safety or ease of operation) may be assessed subjectively; others, such as rate of work, drawbar pull, fuel consumption, etc. may be measured, with varying degrees of accuracy. (Note that animal draught performance can be measured using similar techniques.)

In the example given in Section 11.3.1 it would be necessary to measure drawbar pull and rate of work. It may also be useful to measure fuel consumption and perhaps wheel slip. The point to be made here is that all the factors weighted as 'very important' should be measured, to as high a degree of accuracy as possible. The less important ones should be measured whenever possible, or estimated if not. The test procedure required for drawbar pull, as an example, is normally to use the test tractor to pull a load with a measuring device (dynamometer) in between. Rate of work can be found by measuring the area of operation and dividing by the time taken. Fuel consumption per unit area is assessed by measuring the quantity of fuel consumed and dividing it by the area covered. All these are *procedures*. The question of whether to use a hydraulic dynamometer or a spring balance to measure pull, a flow meter or a gallon can to measure fuel consumption, depends on the equipment available and the accuracy required. This is now discussed.

11.3.4 Measuring equipment and techniques

Two major factors should be kept in mind at this stage. The first is that it is far better to make long-term, realistic tests using rather approximate measuring equipment, than to perform short-term, very accurate tests and extrapolate them into predicted long-term performance. The second factor is that it is surprising how much measuring equipment can be improvised, once the need for it is seen. More of this later.

Returning to the question of extrapolation of short-term tests, a couple of examples should make this point. A tractor is tested for 2 hours while performing a ploughing operation. During this time it is found to have covered say $\frac{1}{4}$ ha. The 'predicted' field performance during ploughing in these conditions is therefore taken as 0.125 ha/h, or 1 ha per day, or 5 ha per week, or 20 ha per month.

In practice it may be found that in 1 month the tractor will only cover 12 ha. Why is this? There may be a combination of reasons, including time lost due to travel, fuelling, adjustment, maintenance, repair, operator fatigue, untypical conditions or obstructions. Had the test been carried out over say 2 days rather than 2 hours, at least some of these factors would have come to light. The rate of work could be measured by pacing or even by estimation, but would still be nearer to the actual long-term rate than that obtained by accurately measuring a short-term test.

Similar remarks apply to the measurement of fuel consumption. Accurate readings of flow rate through a flow meter, when extrapolated, are not likely to give a very realistic total for fuel consumption over a period. The differences could be due to spillage, pilfering, high (or low) consumption in untypical conditions, and so on. Again it will be more realistic to take less precise readings over a longer period. Measurement techniques for this and other parameters are shown in brief in Table 11.3. Some useful details are as follows:

TABLE 11.3 Measurement Techniques for Performance Factors

Factor	Equipment	Techniques
1. Drawbar pull (kN)	Dynamometer	Load vehicle/skid/tine (chisel plough)
2. Distance (m)	Tape or metre stick	Direct measure (or calibrated pacing)
3. Time (s or h)	Stopwatch or hourmeter	Charts or daily totals
4. Slip (%)	Chalk	Mark tyre and count revolutions
5. Fuel (l)	Flowmeter or can/spring balance	Weigh can before and after filling machine
6. Draught force (kN)	Towing vehicle + dynamometer	(a) Tow machine + implement (b) Tow machine (c) Force is (b) − (a)

Derived factor	Combination of:
7. Drawbar power (kN)	1 and 2 and 3
8. Speeds (m/s)	2 and 3
9. Rate of work (h/ha)	2 and 3
10. Fuel/unit area (l/ha)	5 and 2 and 9

Other factors		(Subjective) technique
11. Safety	−	Accident record or survey
12. Ease of operation	Oxygen uptake/ heart rate	Observation/survey/ergonomic study
13. Manoevrability	Tape	Measure turn round distance/area
14. Repairs/downtime	−	Records
15. Status/quality		Survey/observation

(i) Fuel consumption. Over a period fuel consumption may be measured to a satisfactory accuracy by weighing the fuel can before and after filling the tank (using a small spring balance). The specific gravity of diesel fuel may be taken as 0.82, and petrol as 0.76, giving 1.2 l/kg for diesel fuel and 1.3 l/kg for petrol. By recording the total hours worked the fuel consumption per hour may be found; similarly a more useful figure—fuel consumption per unit area while performing a given task—may be derived by measuring cultivated area.

(ii) Area cultivated. The area cultivated may be measured by tape but this is not usually necessary. Sufficient accuracy can usually be achieved by calibrating your pace (take 10 steps and measure the distance covered, say 9 metres; your step is thus 0.9 m in length) and then pacing the field and multiplying by the calibration factor. If you can adopt a metre-length pace for this purpose, so much the better. Slightly improved accuracy can be achieved by bending 2 m of light pipe into a V measuring 1 m across (Figure 11.1). By swinging the resultant pace measurer around on to alternate ends as you walk, it will mark off the metres very rapidly.

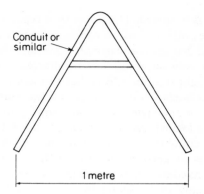

Conduit or
similar

1 metre

FIGURE 11.1 Device for measuring
small field size

(iii) Time. Seconds can of course be measured on a watch or a stopwatch but are not useful except for short-term readings such as speed measurements while obtaining drawbar power values. *Hours* are more relevant to realistic machinery testing. A convenient way of obtaining these values for field work (particularly if the machine is not under constant supervision) is by the use of hour meters.

Various types are available, including engine-driven electronic versions. Perhaps the most useful for simple engine-powered machinery are the types which operate on vibration. These may be positioned on the machine wherever convenient and will either register elapsed hours on a digital display, or will record on a 12 hour clockwork-driven chart the precise periods when the machine was operating. Both these types are normally tamperproof. The chart type is less convenient to set, but is useful when an operator is working without supervision, as he can later be invited to explain the reason for his $3\frac{1}{2}$ hour lunch break. Since it is enclosed, it is also not too obvious to the operator (as it would be with the digital display type) that the hour meters will unfortunately continue to record while the engine (and perhaps the operator) is idling, even though the machine is not working.

These (slightly cynical) remarks underline a real problem, which is that the results of long-term testing with lower levels of supervision are more likely to be sensitive to operator reliability, and steps may need to be taken to reduce the likelihood of these problems occurring.

(iv) Drawbar pull and implement draught. The pull *requirement*, for example the draught force from a tractor-mounted implement in given soil conditions, can be obtained by pulling the implement through the soil while it is attached to a tractor, both tractor and implement being pulled by a second tractor with a dynamometer in between. The tractor on which the implement is mounted should be in neutral (engine idling or switched off). This gives the total force required to pull the implement, plus that to overcome the rolling resistance of the tractor on

which the implement is mounted. The test is now repeated with the implement raised, which gives the pull required for rolling resistance alone. Subtracting the second result from the first gives a reasonably accurate value of implement pull required. The rolling resistance will not be precisely the same for the two conditions, due to weight transfer effects, but this can usually be ignored.

To obtain the pull *available* from a tractor, it is necessary to pull a given load, again measured by a dynamometer. The load can be another tractor (using the brakes as a load), or a loaded trailer or skid. The advantage of using a second tractor is that, if its operator can see the dynamometer reading, he can adjust the load, using the brakes to keep it constant as the test proceeds. A hydraulic dynamometer able to handle large pulls is expensive and may not be available. A spring balance is usually to be found in most workshops but the range of operation is not likely to be high enough for any except small machines. This problem can easily be overcome by making up a simple lever system (Figure 11.2). By attaching the spring balance to the end of the lever and applying the load at an intermediate point, the effective range of the spring balance is increased, by up to a factor of 10 depending on the point of load attachment. Here the moments about the pivot point M can be seen to be

$$P \times a = R \times b . \cdot . P = \frac{R \times b}{a}$$

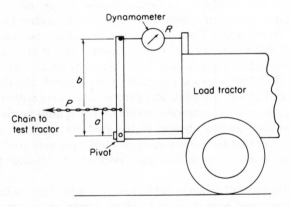

FIGURE 11.2 Converting a small dynamometer into a
useful load measurer

Thus a relatively light spring balance reading to say 5 kN could be made into a dynamometer capable of registering up to 50 kN, simply by making distance $a = 100$ mm and $b = 1000$ mm. A damper can be fitted to reduce oscillations, and the resulting equipment should give useful results.

(v) Slip and drawbar power. Slip values and speed are usually measured during such tests. Slip is easy to measure; the tractor is driven with a raised implement for

10 or 20 wheel revolutions (mark the tyre with chalk) and the distance measured. While under load, the same number of revolutions is counted and the distance measured. It will be found to be lower, usually by a factor of 10–20 per cent. This represents the wheelslip, which can be found from the equation

$$\text{Slip } S = \frac{\text{no load distance} - \text{loaded distance}}{\text{no load distance}}$$

For example, if the first measurement gave 31 m and the second 26 m, the slip is

$$S = \frac{31 - 26}{31}$$

$$= 0.16 \text{ or } 16 \text{ per cent}$$

By measuring slip during a series of tests, each with a higher load, a slip-pull graph may later be drawn. It will be of the form shown in Figure 11.3. This gives the pull which the tractor can develop at various slips.

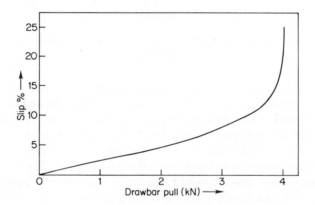

FIGURE 11.3 Pull/slip curve for a small tractor

Sustained slip values of above 15 per cent are usually considered uneconomic. If speed measurements (distance divided by time) are also taken it is then possible to construct curves of drawbar power against pull for each gear. Drawbar power is simply pull times speed: as the pull is increased, so the drawbar power increases, but there will come a point when the increasing slip (which reduces the forward speed) acts to start reducing drawbar power. The curves therefore reach maxima, as shown in Figure 11.4, after which the tractor will stall in the higher gears, and 'spin out' (100 per cent slip) in the lower ones.

11.4 ASSESSMENT

When the required tests have been carried out to a satisfactory level of accuracy with the alternative machines being considered, a set of test results will then be

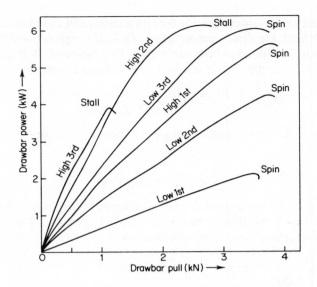

FIGURE 11.4 Drawbar power/drawbar pull curves for a
small tractor

available which may be used to make an *assessment* of the machines concerned.
Certain derived results can be obtained from the raw data. In the previous section
it was explained how pull plus speed plus slip readings can be used to obtain
drawbar power characteristics. Such results are interesting but, because they
relate more to how *efficient* the machine is, rather than to its *effectiveness*, they
will be less useful than simpler but more task-related figures. The most important
test figures are usually those relating to time and cost. For field equipment, then,
it is *rate of work* (in ha/h) which is important, together with *fuel consumption per
unit area* (in l/ha) for a given task. Knowing the area of a field (ha) it is then
possible to predict realistically (i) how long it will take to prepare the field using a
given machine; and (ii) how much fuel will be consumed while doing so.

For other operations such as processing, the figures will normally be related to
quantity (mass or volume). The rate of work then would appear as kg/h or m³/h,
and the fuel consumption as l/kg or l/m³. The great advantage of producing test
results in units such as these is that the performances of very different machines
can be compared. This cannot be done if more 'basic' figures are given (such as
fuel consumption per hour, where a small tractor may appear much more
attractive than a large one, until it is realized that the latter will work much faster
and the fuel consumed per hectare may then be a good deal lower). There is a
temptation, particularly with manufacturers, to specify such basic figures, but
these should be discarded in favour of the work-based results discussed above.

A convenient and effective way of using test results is to compare them with the
characteristics required, which were discussed in Section 11.3.2. It will be
remembered that each requirement was subjected to a 'weighting' procedure, and

this will be applied at the end of the next stage. First it is necessary to 'rate' the machines against each requirement, giving a mark out of 5 for each one (where 5 is excellent and 1 is very bad). This may be most easily done using a matrix, as shown in Table 11.4.

TABLE 11.4 Unweighted Evaluation of Alternatives

Requirements	A	B	C	D	Total
Machine 1	3	4	1	2	10
Machine 2	2	1	4	3	10

In the example given it may be seen that each machine does rather well for two (different) requirements and rather badly for the other two requirements. The total marks would be 10 in each case (out of a total of 20, making 50 per cent). At this stage, however, we are treating all requirements as being of the same importance, whereas we found in Section 11.3.2 that they will in fact have different weightings. Let us now apply the weightings so that the relative importance of the various requirements is reflected in the results.

It will be recalled that the weightings given to each requirement in the example in Section 11.3.2 were:

Requirement	A	B	C	D
Weighting	5	5	1	1

We now apply these weightings to the matrix, by multiplying the appropriate factor by its weighting. Thus the matrix now appears as Table 11.5. Machine 1 can now be seen to be far more appropriate *for the tasks required* than Machine 2. (The maximum possible marks total 60, thus machine 1 has a score of 63 per cent and machine 2 has 37 per cent.)

TABLE 11.5 Weighted Evaluation of Alternatives

Requirements	$A \times 5$	$B \times 5$	$C \times 1$	$D \times 1$	Total
Machine 1	3×5 $= 15$	4×5 $= 20$	1×1 $= 1$	2×1 $= 2$	38
Machine 2	2×5 $= 10$	1×5 $= 5$	4×1 $= 4$	3×1 $= 3$	22

This procedure, although simple, is very useful as it can be extended to cover any reasonable number of requirements (say 10), and also a larger number of machines. It will draw attention to those machines which are more promising than the rest, and does so in a way which is relatively free from bias. It is important, however, to apply the weighting procedure, preferably as the last stage—the results of not doing so can be seen from Table 11.4, where both machines ended up with the same score. When the relative importance of the requirements was introduced, by means of the weightings as in Table 11.5, one machine was seen to be far more appropriate than the other.

11.5 IMPROVING PERFORMANCE BY MODIFICATION

A machine is usually made available with a certain specification and consequently a certain performance will result. Sometimes it is possible to select a different specification, but this is not usually the case. In most situations any modifications will have to be done locally and the evaluation procedure described in previous sections can then be of considerable assistance. Let us take the case of machine 2, for example, which was found not to perform well in two important areas. If this performance could be raised to say three out of five in each case, it would score almost as much as machine 1 and would then be a viable alternative. The matrix in fact draws attention to important areas where performance is not satisfactory.

Can anything be done to modify the machine so as to improve the performance? Certainly there are possibilities, and they are more likely to be realized where a machine is designed on the 'chassis plus components' basis rather than as a complete unit like a large tractor. Fortunately many small machines are of the former type, and there are then distinct possibilities for useful modifications. Some are described in the following section.

11.5.1 Engine power and type

Small engines usually rotate at about 3000 revs/min and the output is normally from a shaft about 25 mm in diameter rotating anti-clockwise. Where the drive then proceeds via belts or chains it is usually quite easy to arrange for an alternative engine to be substituted, by modifying the engine mounting plate. If the output shaft is higher than previously the engine will need to be moved nearer the next stage in the drive line, in order to keep the centre distance the same (Figure 11.5). (The output shaft should be located on the curve AB, centred at O.)

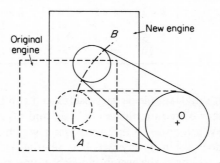

FIGURE 11.5 Repositioning a belt drive due to an engine change

Some engines provide drives from the flywheel end—normally clockwise rotation—as an alternative option, but others rotate in the opposite sense. When changing engines, therefore, always check the rotation.

Changing the *type* of engine (say from petrol to diesel) in order to improve fuel consumption is usually possible, provided engine cooling, exhaust, and vibration effects are catered for. Changing the engine *power* (usually upwards) may put more load on the transmission, so that must be assessed in order to decide whether it is likely to accept a higher power throughout. Chains, shafts, and gears will normally be oversized to a certain extent, but it should be realized that increasing engine power significantly by say a factor of 1.5 or 2 is likely to result in earlier failure or wearout of components. This may be regarded as acceptable in some applications, particularly when the extra power is not likely to be used at low speeds (for example because of traction limitations).

11.5.2 Wheel and tyre size

This factor is, as discussed in Chapters 6 and 9, an important element in tractive performance (drawbar pull, slip, rolling resistance, and mobility). It is also a major determinant in machine forward speed, so fitting larger diameter wheels/tyres is an easy way of making a machine go faster. Finally, ground clearance will be increased with larger diameter wheels or tyres. Either or both can be changed—larger tyres can sometimes be fitted to given wheels (eg. 7.50 −

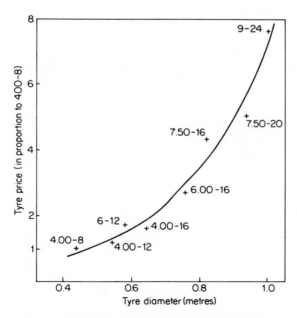

FIGURE 11.6 Tyre price as a function of diameter

16 instead of 6.00–16); larger wheels of course imply larger diameter tyres as well, but not necessarily wider ones (eg. 7.50–20 instead of 7.50–16).

When tyre and/or wheel sizes are increased it will have a considerable affect on traction performance *if* the machine was undertyred in the first place. For example, a heavy small tractor (over 1 tonne) with tyres smaller than 6.00–16 will probably benefit considerably from increased traction and reduced rolling resistance if fitted with larger wheels (say 20 inch), whereas a light single-axle tractor fitted with say 6 = 12 tyres (equivalent to 6.00–12) will not benefit much from an increase in tyre size because of the low machine weight.

It should be noted, also, that tyre prices tend to increase almost exponentially with size (Figure 11.6). The fitment of larger tyres will therefore be likely to increase machine cost, unless certain tyres are already freely available in the country for other applications, in which case they may be sold at an acceptable price.

REFERENCES

Dwyer, M. J., Evernden, D. W., and McAllister, M. (1976). Handbook of agricultural tyre performance, NIAE, Silsoe, Report No. 18.

Inns, F. M. (1978). Operational aspects of tractor use in developing countries, *The Agricultural Engineer (UK)*, **33**, 2, 52–5.

Pollard, S., and Morris, J. (1978). Economic aspects of the introduction of small tractors in developing countries, *The Agricultural Engineer (UK)*, **33**, 2, 29–31.

12 *Machine Maintenance and Local Production*

Ox carts can be produced locally using some proprietary components

12.1 MAINTENANCE

Machines, being inanimate objects, have no power of regeneration and so whenever machines are used it is essential to provide a maintenance service consisting of spare parts and trained personnel, otherwise the effective operating life will be very short.

The machine is designed to do a specific task at a particular price, and under average operating conditions will continue to do this satisfactorily for a number of hours life. For instance, a medium-sized agricultural tractor is designed for a 10,000 hour life provided certain parts are replaced regularly and provided parts are not allowed to degrade (e.g. through corrosion); broken parts are replaced before further damage is caused; oil and air filters and oil are changed at specified intervals, etc.

During the tractor's life it is expected that some parts will be replaced a number of times. For instance a set of tyres may last only two seasons or 2000 hours under the particular operating conditions. The major mechanical parts, however, should work reliably for the whole length of time. The tractor can, of course, be kept going much longer than this but major replacements become necessary, such as entire engines, gear boxes, final drives. This becomes progressively more uneconomic as the tractor spends more time in the workshop and less on productive work. By this time, also, the tractor is a number of years old and additional problems of degradation of the plastic components exposed to the sun, and general fatigue problems, such as cracks and broken welds in the main structure, become important, causing at the least inconvenience and at most, major breakdowns.

A further consideration is the availability of spares from the manufacturer. He may produce the same model for some two or three years, then update and improve some of the parts to give perhaps a larger load-carrying capacity, more speed, more power, or a change of material to give an easier production or cheaper cost of a particular component. These changes are made, for many reasons, typically to give a better product, improve the manufacturer's profitability, widen the machine's application and operating conditions, and keep ahead of the competition. By the time the tractor is, say, ten-years-old there will have been many improvements and changes, so it is no longer economic for the manufacturer to keep supplying out-of-date parts and hence it will become progressively more difficult to keep an old machine working, because of the increasing difficulty of finding the spare parts.

It is for these reasons that the machine has an economic life as well as a mechanical one and it is necessary for the farmer to change the machine for a new one at regular intervals in order to obtain machine efficiency and maintain the effectiveness of his investment.

If the operating conditions are more severe in a particular area than those the manufacturer considered as average, then the machine components will fail in a shorter time. This will result in higher running costs in terms of lost production and increased spares cost. The machine will become uneconomic if the conditions are too severe for it, in which case a heavier duty machine should replace it. This will in turn cost more and the decision to do this will depend on the economics of the operation, such as the level of profitability required before the project is abandoned.

The relationship between load and life can most easily be illustrated by taking ball bearings as the example. The bearing rotating under load will eventually fail, usually by surface fatigue of the rolling elements. The manufacturer will test samples of bearings and from the results publish a Basic Dynamic Capacity, C, that is the load that the bearing can carry for one million revolutions. This is known as the B_{10} life as 10 per cent of a sample of bearings tested in this way will fail before they reach 1 million revolutions. If the required life is more than 1 million revolutions a table of factors, or nomograph, can be used to correct the load-carrying capacity (in this case reduce it).

In the nomograph shown (Figure 12.1), the radial load required to be carried, called the Equivalent load, P, is calculated from the requirements of the particular piece of equipment. The calculation combines the radial and axial

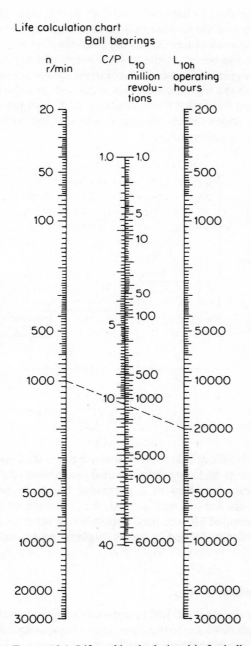

FIGURE 12.1 Life and load relationship for ball bearings (after SKF 3200 E, reproduced by permission of SKF)

loads into an equivalent radial load. The C/P ratio is read off and this factor is multiplied by P to give the C value that is required of the bearing. The C values are listed by the manufacturer for all the bearings that are available.

The information is based on the basic theories of fatigue failure which are summarized in the graph in Figure 12.2 as a stress against number of cycles curve. The curve shows that a high load, S', can be carried for a short time, N', while a lower load, S'', can be carried for a longer time, N'', confirming the fact that a machine worked under severe conditions will not last as long as a machine worked under less severe conditions.

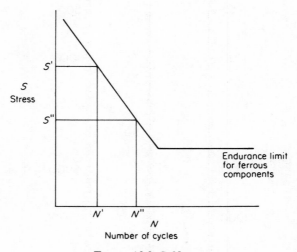

FIGURE 12.2 S–N curve

It should be pointed out that these results are based on a large number of tests on many components and state the statistical probability of a certain life before failure of significant numbers of a particular batch. In practice, where the operating conditions are less well-defined, the actual life of the machine is less certain. This concept of life and load applies to all other machine components which are dynamically loaded, such as engines, gears, axles, pumps, pipes, frameworks, etc.

12.1.1 Spares and servicing

The reasons why the machine will require servicing and spare parts have been outlined. Unfortunately, due to the wide range of operating conditions and many differing types of machines used, it is difficult to be very precise when it comes to making recommendations of the types and numbers of spare parts that will be required during the machine's life. The most practical way of arriving at a suitable recommendation is to base it on carefully kept records over many operating seasons of similar equipment working a similar system in a similar environment.

The manufacturer of the equipment can provide a recommended list of parts that would be required for the particular circumstances (to keep the machine working for the desired length of time), based on his previous experience. When selecting machines from competing manufacturers it is important to ensure that this recommended list is complete and a reasonable estimate, as it will have a large effect on the estimated maintenance costs; a short list implying that the machine does not need spare parts is likely to be very misleading.

The term 'maintenance' can be defined as: 'A combination of any actions carried out to retain an item in, or restore it to, an acceptable condition'. Clearly, from the previous discussion the reasons for the machine requiring maintenance are very complex as there are many interacting factors. In recent years a new term has come into use which attempts to bring all the aspects of maintenance under one heading—terotechnology.

Terotechnology has been defined as:

A combination of management, financial, engineering and other practices applied to physical assets in pursuit of economic life cycle costs; it is concerned with the specification and design for reliability and maintenance of plant, machinery, equipment, buildings and structures, with their installation, commissioning, maintenance, modification and replacement, and with feed back of information on design, performance and costs. (Department of Industry, 1975)

The principal objectives can be defined as:

(1) To extend the useful life of assets.
(2) To ensure availability and readiness of the system.
(3) To obtain maximum return on the investment.
(4) To ensure safety of personnel and bystanders who use or come into contact with equipment.

Maintenance can be either planned or unplanned.

Unplanned or emergency maintenance is the result of a malfunction of the equipment that was not anticipated. As a result the machine is inoperative and the production system of which it is a part cannot function as originally planned. This causes the maximum disruption of production and can result in loss of product and serious financial loss. Obviously this situation must be avoided if at all possible; this can be done by using planned maintenance. This is normally split into two activities; preventive maintenance and corrective maintenance.

Preventive maintenance is carried out at pre-determined intervals, according to prescribed criteria, and is intended to reduce the likelihood of an item not meeting an acceptable condition. An example of this is regular changing of oil filters after some time in service but before they cease to function as filters, by becoming blocked up, as this will lead to a more serious and costly failure of the engine.

Corrective maintenance is carried out to restore (including adjustment and repair) an item which has ceased to meet an acceptable condition. An example of this is replacing the main clutch assembly when there are signs of malfunction but before it has ceased to function altogether. For instance, if it starts to judder or the thrust bearing squeaks—the look, listen, and feel approach. As many machine components function perfectly for some time and then show a more or less rapid decline in performance, there is often sufficient warning to change the component at a convenient time before the emergency situation arises.

For the larger, more expensive pieces of equipment it is often worth introducing a formal checking procedure which can give information on when certain parts require changing. For instance, on large earth-moving plant the engine and gear box oils can be sampled for metal contamination and if, say, the phosphor-bronze content suddenly increases this will indicate that certain components have developed a fault; these can then be replaced before other more serious faults develop. The noise emission of ball and roller bearings can be monitored—an increase in noise level will indicate that the bearing is nearing the end of its useful life. The exhaust temperature at each exhaust valve can be monitored; any sudden increase will indicate that the valves are not sealing properly and either they need regrinding or they and their seats require replacing. This can be done before the engine loses too much power and other more serious faults like cracked heads have occurred.

The decision to adopt these more elaborate systems to control the maintenance schedules for the equipment will depend on the overall economies of the operation and should be made by the managers of the project when all the facts are known. For the small-scale mechanization system and the small farmer, such elaborate systems are not feasible but the same principles can be used—the look, listen, and feel approach. For this to work, it is essential that the mechanic has adequate training and experience at the level required for the machinery being used. Keen observation and a sympathy for the job are also important requirements; the man who notices a slight wobble on a wheel can tighten the wheel nuts before they strip off or change the wheel bearing before the casing and shaft are seriously damaged; a few moments work can save many long expensive breakdowns.

12.1.2 The Farm workshop

In order to carry out the required maintenance, a suitable workshop and tools must be provided. The sizes and complexity of the building and tools will depend on a number of factors:

(1) The size of the farm.
(2) The number of pieces of equipment.
(3) Proximity to the main dealer or manufacturers of the equipment.
(4) Ease of transport of damaged machinery or of spare parts to the workshop or dealer.
(5) The degree of training and competence of staff.

This book is primarily about small farm mechanization, which limits the types of systems that are appropriate. The system adopted will also be influenced by the overall social and economic system in the particular area, whether it is a government-organized system or a private entrepreneur system, and by the sources of funds for investment in the workshop, the general level of incomes of the farmers and their ability to pay for the service. We assume that the farmer will have the minimum equipment, tools, and knowledge to operate his farm. This will mean that he can carry out only routine servicing of engines and component changing of the parts subject to wear such as soil-engaging tips to implements.

12.1.3 Small farm facilities

The following list summarizes the requirements:

(1) *A raised hard standing*: to give good drainage with a simple pole structure and roof (grass) to give reasonable protection from sun and rain. The size of the structure should be adequate to store all the equipment the farmer has and to provide sufficient space to be able to walk round each piece of equipment to carry out maintenance work.

(2) *Spare parts store area*: lockable for adequate security.
Parts required:

  (a) engine oil filter element and seal ring;
  (b) fuel filter element and seal ring;
  (c) air filter element and seal ring;
  (d) spark plugs and points if petrol engine;
  (e) spare V belts;
  (f) spare control cable (throttle and engine stop);
  (g) spare nuts and bolts;
  (h) spare soil-engaging parts for implements.
  (i) stock parts for one service only, replace immediately after service.

(3) *Fuel and lubrication store*: steel containers are the most suitable. The most convenient fuel containers are a number of 200 litre drums, with taps mounted on the sides at a height sufficient to give gravity fill of the machines' fuel tanks.

(4) *Tools*:

  (a) hammer (1 kg);
  (b) cold chisel;
  (c) open-ended adjustable spanner—small;
  (d) open-ended adjustable spanner—large;
  (e) screwdrivers—small and large;
  (f) screwdriver—cross-drive type—medium;
  (g) pliers;
  (h) hacksaw and high-speed blades;
  (i) flat file, round file.

(5) *Instruction manuals*: training for the farmer to learn correct operating and maintenance procedure for his equipment.

12.1.4 District workshop facilities

To back up the farmer with the system previously outlined, it will be necessary to provide a more comprehensive workshop and spare parts service at the village level or at the local area level depending on the size of the mechanization system adopted in the locality. Typical requirements are summarized below.

(i) The site. This should be well-drained and easily accessible to the area, for example with suitable roads or all-weather tracks, and surrounded by a security fence. The open-air machinery storage area should be arranged so that equipment can be parked in rows side by side with sufficiently wide spaces between rows so that the equipment can be driven directly in and out without having to do complex manoeuvres. A loading ramp with a vertical wall at truck-bed height in an easily accessible place is also essential.

There should also be a covered machinery storage area. It will be advisable to provide covered accommodation for the more expensive, larger, seasonal equipment, to give protection from sun and rain to minimize general degradation during the greater part of the year when the machinery is not being used. The floor area should be such that it is possible to walk round the machinery easily (minimum 600 mm space). The area should be large enough to take about 15 per cent of the total number of machines from the scheme or locality in addition to the permanently stored machines. The 15 per cent represents the proportion of machines that will probably be broken-down at any one time and be in the workshop for maintenance. Under good conditions the figure is likely to be smaller while under poor conditions it is likely to be larger.

(ii) The workshop building. From the number of machines, and taking the 15 per cent figure for breakdowns, the floor space taken up by this number can be calculated from the average size of the equipment. It should be noted that the size of the individual bays of the building should be large enough for the largest machine and the doorway tall enough. One-third of the bays should each have a pit, for which Figure 12.3 shows a typical type. Note it has easy-access down steps and a slightly sloping floor with a sump at the low end, big enough to bale out oil and water with a 10 litre bucket.

The general construction of the building will not be discussed here but it should have the following characteristics:

(a) good, well-drained site;
(b) high-quality concrete floor;
(c) strong fireproof construction;
(d) good ventilation—probably open eaves covered with wire mesh (for security and to keep out animals);

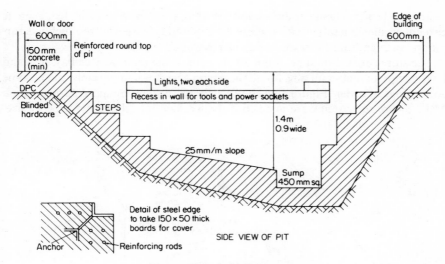

FIGURE 12.3 A workshop pit

(e) good lighting, especially in the pit and over the work benches;
(f) floodlights in the yard.
(g) an electrical, water, and air ring-main so that the equipment can be plugged in near the job to avoid dangerous trailing wires;
(h) a vehicle wash area outside with drainage away from the building, which will help to keep the workshop area cleaner and avoid dirt contamination of the repairs.

When selecting the size of the electrical and air ring-main it is reasonable to assume that half the available equipment will be used at any one time. Electric welders should be wired separately and the system sized to allow all to work at the same thing. *Note*—air line and air tools are usually only justifiable for the larger scheme.

An overhead crane should be provided and should preferably be a free-standing gantry type on wheels capable of lifting 2 tonnes and able to straddle and clear the largest piece of equipment that is likely to require servicing. At least two small and two large slings should be provided and there should be one crane for every five bays. The advantage of this type of crane is that it is independent of the building which may be of local materials and of unknown strength and it can be used outside in the yard for the extraordinary job.

Each bay should have a bench fitted with a heavy-duty 125 mm vice (125 mm jaws). The bench should be attached firmly to the floor and may have storage cupboards underneath.

For each five-bay unit there should be an additional welding bay with screens or walls to protect other workers from the welding flash. Additional fire extinguishers should be available in this area. A dry storage area for the welding rods is essential.

234

For the larger scheme additional specialized areas may be provided, such as carpentry; sheet metal; machine shop; battery and tyre; specialized fuel injector, pump, and ignition; lubrication bay (air-powered type).

The stores area should be about 20 per cent of the main service bay area and have adequate racks and bins for all the parts to be stored separately and in some sort of logical order. Figure 12.4 shows a simple system for spare parts control. More elaborate systems are justified if the workshop is above the ten-bay size.

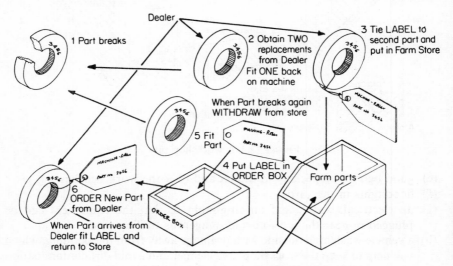

FIGURE 12.4 Spare parts control (after ADAS, 1978)

Use LABELS for large parts singly and for whole bins of small parts and quick-moving items such as plough shares. Make a firm rule that when a bin gets down to the LAST SIX items the LABEL is put into the ORDER BOX

(iii) Fuel and lubrication store. The size of the store can be calculated from the number and size of the tractors in the scheme and the required frequency of resupply. It is often convenient to size the fuel tank to take the tanker component load or the whole tanker load, that is, so that the storage tank is 20 per cent larger to ensure the full load can always be accommodated. All tanks should have facilities for dewatering and desludging.

(iv) Tools. A complete list of the tools required appears in the Appendix.

(v) Labour requirement in workshop. One fully trained mechanic can be expected to look after ten tractors and their associated equipment by himself, or perhaps with an unskilled assistant. For every five mechanics there should be one fully qualified welder and storeman. At about this size of operation a full-time manager will become necessary in order to co-ordinate the work for maximum efficiency.

(vi) Records. Good records of the machines and repairs carried out are essential if the system is to work reliably; the information can also be used for adequate financial control, which will allow the proper management decisions to be taken.

Figure 12.5 shows a typical planning chart for maintenance. Each machine should have a record card; Figure 12.6 shows a typical example. When the machine is taken to the workshop it should have a workshop job card which tells the mechanic what to do, and records the parts used and the time taken.

Each machine should be provided with an instruction book, parts list, and service manual. These should be available to the mechanic so that he can ensure the right parts are fitted in the correct way. The documents should be returned to the store after the job is complete.

(vii) Safety equipment. The workshop should have the following equipment in good working order:

Fire extinguishers—dry powder and carbon dioxide for electrical and liquid fires; foam extinguishers—for burning liquids and solids *not* electrical fires; soda acid for wood/paper fires only; gas pressure—water-based so only for solid burning materials; fire buckets—sand-filled; fire blanket.
First aid kit.
Clothes (*protective*); goggles; gloves; breathing masks for use when paint spraying; ear defenders; specialist safety equipment.

(viii) Safety instructions. Make sure that at least one of the staff on duty each shift has basic first aid training. Everyone should know where the emergency power isolation switch and fuel turn-off valves, etc. are situated and always wear the correct clothes for the job, not loose flapping garments which can get caught in moving parts. No-one should ever enter a closed space, for example a grain silo or large storage tank, without an assistant outside. Everyone should use safety harness when appropriate, tie ladders, use rails on scaffolding, and not walk on roofs unaided.

12.2 LOCAL PRODUCTION

12.2.1 Characteristics

The production of various objects has been undertaken in almost all developing countries for centuries. Examples are pottery, carvings, wooden and sisal products, agricultural implements, ox carts, and so on. These production systems have the advantage of using local materials and locally developed skills, requiring little or no imported materials and being close to the demand side, thereby remaining in touch with the market requirements.

Imported products possess none of the above advantages. They are often expensive and sophisticated, sometimes not suitably developed for local

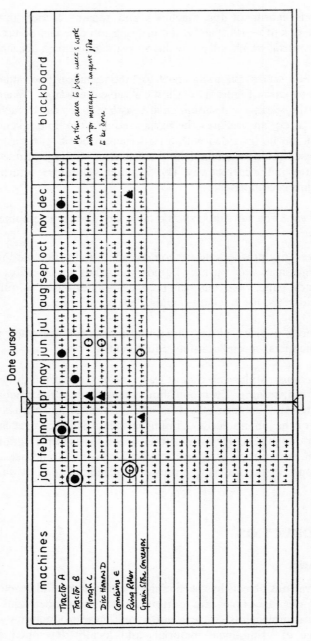

A chart whether home-made or purchased complete is a valuable asset in planning work in the workshop

The main symbols used are :

● – Regular service   ▲ – Visual inspection of repair need   ◯ – Cancelling symbol indicating work carried out

FIGURE 12.5 Maintenance chart (after ADAS, 1978)

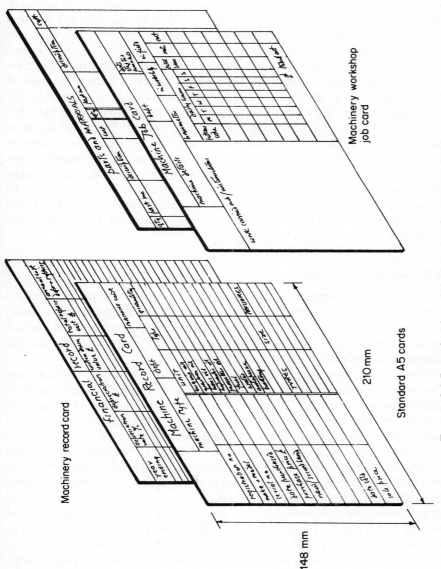

FIGURE 12.6 Record and workshop cards (after ADAS, 1978)

conditions, they are not backed up by a suitable repairs/spares infrastructure and their replacement involves long delays.

The overriding feature that they do possess, however, may be summed up in the word 'interchangeability'. Such components are produced in large numbers, often on complex production and assembly lines, which is the most effective way of ensuring consistency of quality and size. Modern production methods ensure that products are, to all intents and purposes, identical so that not only original parts but also replacement parts will always fit the machine for which they were designed. Because such machinery inevitably consists of a large number of different parts it is essential that such universal consistency is achieved.

Local production from rural industries in developing countries, on the other hand, is usually aimed at simple devices in which it is not important that exact consistency is obtained. Even when replacements are provided (for example a new hoe blade) a different size can nearly always be accommodated by slight modification. (Such a possibility does not exist in cases like, for example, replacement of a diesel engine fuel injection pump, where *all* dimensions must be identical to the original.) Local production is usually carried out in batches, and uniformity within batches even from a single craftsman is by no means guaranteed.

12.2.2 Benefits and problems

It is tempting to attempt, at a stroke, to eliminate the disadvantages of imported products by producing components locally. Foreign exchange is thereby saved, local employment is generated, national pride in production facilities and products is established, spares are more easily and quickly available and the product is geared more closely to local needs.

These potential advantages certainly exist but, for a number of reasons, are often not fully gained. One reason is that demand is usually concentrated on import-substitution rather than on locally designed equipment. Since the imported product (such as a conventional medium-sized agricultural tractor) is made up of complex subassemblies, the usual procedure is to start with assembly-only lines using CKD (completely knocked down) components which are imported instead of finished machines; the aim is then to develop over a period of years the facility to manufacture an increasing proportion of the machine locally. Since the design of the machine is, however, likely to be essentially the same as the one originally imported complete, it is necessary to provide the same kind of complex and expensive production lines which exist in the country of origin. In this case the purchase, installation, and management of the production facility is likely to call upon resources which are already scarce in the developing country concerned. Processes such as that described have been instituted with some degree of success in a number of developing countries and will probably continue to be the only viable way of increasing the local content of complex machinery such as tractors and vehicles.

For less sophisticated machinery there is an enhanced possibility of local production. One reason is that such machinery is not generally produced in developed countries (and even when it is, it will not normally utilize mass production techniques). The design of the equipment, therefore, is more likely to lend itself to local part-manufacture at least, and also to local design modification.

12.2.3 Facilities for production

Let us consider the example of a small tractor. It will usually consist of a power component (engine), transmission components (belts, chains, axle), running gear (wheels), and control systems (hydraulics, electrics, etc.). All these parts will need locating relative to one another so that they can function. This is usually done by means of fabricated chassis. Provided the components are located correctly and the structure is sufficiently stiff and strong, it is not particularly important whether the chassis is constructed of plate, angle, channel, rectangular hollow section, or even round tube. There is thus considerable scope for local redesign to suit available materials and production facilities.

Such facilities need not be very sophisticated. To produce an adequate small tractor chassis it is probably only necessary to have a drill, a power saw, and welding gear. The use of simple welding jigs will often ensure sufficient consistency in dimensions to allow satisfactory production even on a batch system.

The main criterion is that quality must still be controlled. This particularly includes the quality of local steel and other materials, and the integrity of the welding. Even a simple device must be built up to an acceptable quality if its reputation is not to suffer. It will already tend to be regarded as an inferior product compared with more sophisticated imported machines, simply because it *is* less sophisticated. Provided it operates effectively and economically, this problem can perhaps be overcome, but the machine must also remain reliable. Quality control, allied to careful redesign by qualified personnel who understand the engineering reasons for certain component specifications, is an important way of ensuring this continued reliability.

Other aspects of production are as follows. The machines that are selected should be the cheapest available consistent with having adequate performance. Because of the high cost of labour in industrialized countries it is often thought that parts or even complete machines can be made more cheaply by the local people. The situation must be analysed very carefully; all the relevant factors should be considered before a decision is taken.

There are a number of parts that almost certainly cannot be manufactured locally due either to the advanced technology required or to the very large investment required to get the unit cost low enough. Examples are: ball bearings; roller transmission chains; diesel engine injectors and pumps; pistons, shell bearings; gears; electronic components, microchips; crawler track running gear; hard facing rods. There are of course many others.

Many farm machines can, however, be made from standard purchased parts

240

such as bearings, V belts, gears, etc., joined by a suitable framework designed to be the correct shape and to be strong enough for the imposed loads. The 'framework' could be the chassis for a simple trailer or the structure of an oil-extractor machine. One of the most widely adopted methods of manufacturing such 'frameworks', as was mentioned earlier, is the cut and weld method and provided the various steel sections are available locally it is often possible to make the machine at an economic price for the local market. Local industry should be encouraged as it increases the pool of knowledge in the area and thus there is more chance of the machine being used successfully and evolving to suit the local conditions.

This method of manufacture produces very satisfactory results if a few basic procedures are adopted. If the component is highly stressed, weld preparation is

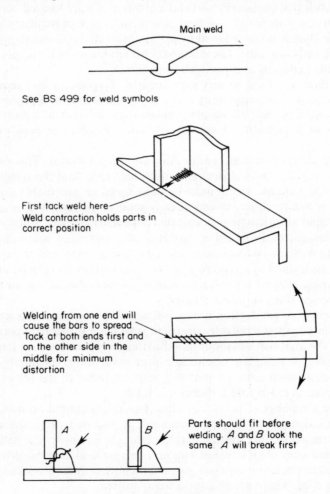

Main weld

See BS 499 for weld symbols

First tack weld here
Weld contraction holds parts in
correct position

Welding from one end will
cause the bars to spread.
Tack at both ends first and
on the other side in the
middle for minimum
distortion

A   *B*   Parts should fit before
welding. *A* and *B* look the
same. *A* will break first

FIGURE 12.7 Cut-and-weld fabrication technique

required in order to obtain full penetration of the weld and consequent high strength. Welding should be done in the correct sequence to avoid or minimize weld heat distortion of the structure, and the worker must make sure the parts fit properly before welding. Slag should not be included in the weld (an experienced welder should not have this problem). The welding should if possible be carried out 'down hand' as better quality welds are more easily obtained than from all-position welding. Wet or damp rods should not be used and the correct voltage and current must be employed as stated by the manufacturers of the rods. For large-quantity production a gas-shielded arc using CO_2 and an automatically controlled plain feederwire gives better quality welds with less operator skill.

Welding jigs should be used to help ensure that all the parts are the correct size. For soil-engaging parts where heat-treated alloy steels are used, it is often possible to redesign the component to use mild steel with a layer of weld-deposited hard facing on the soil-contact surface—here a lattice deposition pattern gives the most economical effect. Figure 12.7 illustrates some of these features.

Many parts of machines can be redesigned to use locally available materials—for example a lot of metal parts can be replaced with wood. There are many types of wood available, some of which have excellent properties, for use in slow-speed bearings for example. Locally made wooden wheels and axles with the minimum of metal parts are used very successfully in some areas, and locally made plywood can be used for highly stressed structures if the correct design procedures are adopted.

Other materials lend themselves to simplified local construction of large structures—for instance, ferrocement construction of pontoons and storage tanks.

Many innovations in design and choice of materials are possible to provide a machine more suitable for local conditions. The people most likely to find these attractive solutions are those actually working there who have a sound engineering education and who can apply basic principles to their situation. Education is the key to progress.

REFERENCES

ADAS (1978). Ministry of Agriculture Fisheries and Food, *The Farm Workshop.*
Department of Industry (1975). *Terotechnology—An introduction to the Management of Physical Resources.*

Appendix: Tools required for a district workshop

Compressed air equipment

Compressor. Complete unit. 2 kW motor.
$0.5 \, \text{m}^3/\text{min}$ compressor. 1000 Pa minimum.
Tank 140 l.
Air pressure hose 30 m lengths. Bayonet quick fittings
Tyre inflator with gauge.
Paint spray gun.
Blow nozzle.

Lifting equipment

2 tonne gantry crane and slings.
1 tonne chain hoist.
Floor jack 5 tonnes.
Bottle jack $2\frac{1}{2}$ tonnes.
2 axle stands $2\frac{1}{2}$ tonnes.
Tow chain 5 m long, shackle at each end.

Cleaning equipment

Water hosepipe in yard wash area.
Paraffin wash tank, manual pump (make from 200 l oil drum).

Lubrication and fuel

Grease pump, high-pressure bucket type.
Hand-lever grease gun.
Hand pump for fitting to oil drums, etc. with 3 m flex pipe and tap on end.
Oil can, pump type 1 l capacity.
Oil measures, 1 l and 4 l capacity.

Welding equipment

Steel table 1.2×2 m top.
G clamps, two off opening to 75 mm.
G clamps, two off opening to 150 mm.
G clamps two off opening to 250 mm.
Electric welder. 250 amps, oil-cooled, on wheels, complete with 30 m mains and 10 m welding cables, earth clamp, electrode holder, full face mask, wire brush, chipping hammer.
Selection of general purpose coated welding rods 2–6 mm diameter for mild steel and alloy steel.

Selection of 'cutting' rods.
Carbon arc attachment } can be used for welding
Set of various sizes of carbons } sheet metal if oxy-acetylene
gas supply is too difficult.

Gas welding set on wheeled frame, bottles with gauges, regulators, flash back arrestor, 10 mm hose. Welding torch, jets 1–5; cutting attachment, jets to cut up to 20 mm thick.
Goggles; heat-insulating gloves.
Soldering iron $\frac{1}{2}$ kg and 1 kg size.
Flux—killed spirits type.
Solder (multicore type).

Mechanic's tool kit

All in heavy-duty lockable steel box:

(1) *Spanners*—open ended;
    miniature or ignition type;
    Standard type 15° offset chrome alloy forging, set of: AF 3/8 to 1$\frac{1}{4}$ inch; BSW 3/16 to 7/8 inch; Metric 8 to 36 mm.

(2) *Ring type*—ditto

(3) *Socket set*—$\frac{1}{2}$ inch drive chrome alloy.
    Drive tools, ratchet reversible;
            speed handle;
            T handle;
            universal joint;
            extension bar—long;
            extension bar—short;
            screwdriver sockets, 12 × 20 blades;
            screwdriver sockets, cross-drive.
            spark plug socket;
            sockets, set of: AF 3/8 to 1 inch; BSW 1/8 to $\frac{3}{4}$ inch; Metric 10 to 30 mm.

(4) *Hexagon wrenches*—(Allen key), set of:
            inch 3/32 to 3/8 inch;
            metric 4 to 10 mm.

(5) *Adjustable spanner*—15° single end chrome alloy;
            150 mm jaw opening 20 mm;
            300 mm jaw opening 35 mm.

(6) *Pipe spanner*—(Stillson) 300 mm size.

(7) *Brass drifts*—
            10 mm diameter, 150 mm long;
            16 mm diameter, 150 mm long.

(8) *Pry bar, punches, chisels*—Alloy steel:

              pry bar 450 mm long. 16 mm diameter;
              punch;
              pin punch—small;
              pin punch—medium;
              pin punch—large;
              centre punch;
              scriber;
              flat chisel 200 mm long, 12 mm edge;
              flat chisel 200 mm long, 20 mm edge.

(9) *Pliers*—

        150 mm self-grip pliers;
        150 mm needle-nose pliers;
        Tin snips, 250 mm;
        Internal and external circlip pliers.

(10) *Scraper*—200 mm.

(11) *Screwdriver*—

        blade 100 mm long, 6 wide;
              200 mm long, 10 wide;
              250 mm long, 12 wide;
        electricians' 75 mm long, 3 wide;
        cross-slot (Philips sizes 1 to 4).

(12) *Hammer*—

        ball pein $\frac{1}{2}$ kg;
        ball pein 1 kg.

(13) *Files with wooden handles*—

    flat, 250 mm long, second cut;
    round, 200 mm long, second cut;
    triangular, 150 mm long, smooth;
    contact file 140 mm long;
    thread file American thread type;
    thread file British thread type;
    thread file metric thread type.

(14) *Hacksaw and blades*—

    300 mm, pistol grip;
    blades 18 teeth per inch;
    blades 32 teeth per inch.

(15) *Measuring instruments*—

    Stainless steel rule, 300 mm, English and metric;
    3 m flexible tape;
    Feeler gauge, 100 mm blades metric.

Tools held in main store, to be issued for a particular job

(16) *Heavy-duty tools*—purchased especially to suit equipment to be serviced,
such as spanners up to 2 inch AF; 50 mm metric;
large adjustable spanner 450 mm long, 53 mm opening;
Stillson, 450 mm long;
socket set, 3/4″ drive, up to 2 AF; up to 50 metric;
spider type wheel brace.

(17) *Torque spanner*—adjustable:
$$0\text{–}70 \text{ Nm}, \tfrac{1}{2}'' \text{ drive};$$
$$0\text{–}400 \text{ Nm}, \tfrac{3}{4}'' \text{ drive}.$$

(18) *Crow bar*—1.8 m
jemmy bar 20 mm × 600 mm;
nail puller (rammer type).

(19) *Sledgehammer*—2 kg:
raw hide, $\tfrac{1}{2}$ kg;
plastic tip, $\tfrac{1}{2}$ kg.

(20) *Putty knife.*

(21) *Wire brush*—25 mm wires.

(22) *Pipe bender*—hydraulic, hand-operated to suit pipe size on site.

(23) *Taps and dies*—
set of metric up to 24 mm;
pipe taps and dies (up to size required);
other American and English threads to suit equipment.

(24) *Range of files*—types: mill, flat pillar, square, round, $\tfrac{1}{2}$ round, triangular
knife:
sizes 150 mm to 300 mm;
smooth to bastard cut.

(25) *Set of twist drills*—high-speed type:
1–10 mm in 0.1 mm steps;
10–24 mm in 0.5 mm steps.

(26) (a) 6 mm, hand electric drill;
(b) 12 mm, 2-speed, hand electric drill;
(c) 20 mm, 2-speed, hand electric drill;
(d) fitted with bench drill, stand and vice;
(e) fitted with magnetic drill, stand and vice.

(27) *Bench grinder*—150 mm wheel, 3600 RPM.

(28) *Angle grinder*—150 mm wheel;
selection of metal cutting and sanding discs.

(29) *Hand reamers*—adjustable type with pilot:
set 8 mm to 45 mm.

(30) Set of 2-arm sprocket pullers;
set of 3, up to 250 mm.

(31) Set of 3-arm sprocket pullers;
set of 4 up to 500 mm.

(32) 600 mm steel rule—English and metric divisions.

(33) Vernier calipers 250 mm.

(34) Micrometer 0–150 mm set;
internal and external.

(35) Telescoping gauges, set of 5 up to 150 mm.

(36) Dial indicator for measuring engine cylinder bores;
dial test indicator, magnetic base type.

(37) Screw thread pitch gauges—
American;
English;
Metric.

(38) Dividers, calipers, and oddlegs.

(39) Engineers' straight edge, 1.2 m long.

(40) Engineers' square 300 × 175 mm.

(41) Protractor, vernier graduation, leg 150 mm.

(42) Spirit level, 300 mm long.

(43) Tachometer, electronic to 5000 rpm.

(44) Stopwatch—electronic.

(45) Surface plate, 1000 mm × 600 mm min.

(46) V blocks (set).

(47) Height vernier.

(48) Hydraulic press 10 tonnes, hand-operated.

(49) Battery charger, suitable for boost charging and direct engine starting.
Battery filler—plastic non-break type.

(50) Hand-operated tyre bead loosener—
impact type;
set large tyre levers;
tyre hammer—hard rubber replaceable tip type;
tyre pressure gauge;
valve-replacing tool;
vulcanising press (auto temperature control type);
tyre liquid fill kit.

For the larger workshop or when some manufactures of parts or even whole machines are considered then additional equipment is required.

(51) *Drilling machine*—12 mm capacity, bench-mounted.

(52) *Drilling machine*—40 mm capacity:
radial arm type, floor-mounted;
set of drills to 40 mm diameter.

(53) *Lathe*—300 mm swing, 1.2 m bed:
multiple-speed gear box, screw-cutting;
powered cross- and longitudinal-feed type;
milling attachment, 3 jaw and 4 jaw;
face plate, taper-turning attachment;
steady 3 jaw type;
coolant circulation system;
set of carbide-tipped tools, green grit grinding wheel for bench grinder;
set of centre drills, set of clamps, drill dogs, and blocks.

(54) *Shaper*—500 mm stroke, full set of attachments and tools.

(55) *Cylindrical grinder* 1 m centres.

(56) *Surface grinder* 1 m table.

(57) *Cutter grinder*.

(58) Milling machine—powered slide type with head for end mills and slotting
attachment 1 m bed. Complete set of cutters and end mills. Rotary table.

(59) *Band saw* 900 mm square table—
power feed; blade welding attachment for wood and metal.

(60) *Horizontal band saw*—
coolant system, for rounds up to 300 mm.

(61) Sheet metal guillotine for 1.2 m wide, 3 mm thick steel sheet.

(62) Sheet metal bender fully adjustable type for 1.2 m wide and 3 mm thick steel
sheet.

(63) *Anvil* 100 kg size.

(64) *Vehicle body repair tools*—
hydraulic frame and body-straightening set;
dollies;
hammers;
spoons;
mallet;
file holder;
body files.

(65) Electric spot welder, up to 2 mm sheet.

(66) Pop riveting gun, hand-held and air-operated type.

(67) *Specialist equipment for major overhall of tractors*—
valve grinder and seat refacer;
cylinder borer;
crankshaft regrinder;

cylinder head planing machine;
diesel injector and pump rebuild and calibration rig;
electronic engine test set;
exhaust gas analyser;
hydraulic press for track pin turning;
air-powered tyre remover;
tyre and wheel balancer;
tracking equipment;
test equipment for vehicle electrics;
rolling road test dynameter.

(68) *Blacksmith's equipment* may be required for specialist work—
forge, motor-driven variable fan;
anvil;
swage block and stand;
set of blacksmiths' hammers;
setsquare, compass;
set of tongs, hardie;
conical anvil horn;
cold chisel;
hot chisel;
punches for round and square holes;
top and bottom pullers and swages;
set hammer;
flatter;
levelling plate.

(69) *Carpentry Tools*—
wood-working bench, 2 vices set at each end in line;
saws, cross-cut and rip;
planes, jack, finishing, block, multiple adjustable type for beading;
adze;
draw knife;
mallet;
hammers, claw type, tack type;
chisels, set 6 to 25 mm blades;
ratchet brace;
screwdriver bit;
drills and augers;
expansion bit to 75 mm diameter;
countersink bits;
rasps, flat, round, $\frac{1}{2}$ round, cabinet;
rule 2 m;
level;
awl;
marking gauge;

carpenter's square;
bevel gauge;
divider;
screw drivers, set of 4;
pliers;
nail punches;
G clamps in pairs, 3 sets to 200 mm;
2 sash clamps;
hatchet, felling axe;
oilstone, coarse and fine;
electric-powered sanding machine;
universal planing and sawing machine.

Index